Visible Light Communication Based Indoor Localization

Visible Light Communication Based Indoor Localization

Mohsen Kavehrad
Reza Aminikashani

CRC Press
Taylor & Francis Group
Boca Raton London New York

CRC Press is an imprint of the
Taylor & Francis Group, an **informa** business

CRC Press
Taylor & Francis Group
6000 Broken Sound Parkway NW, Suite 300
Boca Raton, FL 33487-2742

First issued in paperback 2023

© 2020 by Taylor & Francis Group, LLC
CRC Press is an imprint of Taylor & Francis Group, an Informa business

No claim to original U.S. Government works

ISBN 13: 978-1-138-61747-6 (hbk)
ISBN 13: 978-1-03-265347-1 (pbk)
ISBN 13: 978-0-429-35580-6 (ebk)

DOI: 10.1201/9780429355806

Dedication

Mohsen Kavehrad:
To my family for their love and support

Reza Aminikashani:
To my wife, Sepideh, and my daughter, Eva. when she was not born yet

Contents

Preface ix

Chapter 1   Introduction and Overview 1

Chapter 2   Fundamentals of Visible Light Communication 5

    2.1   System Model 5
          2.1.1   Transmitter 5
          2.1.2   Receiver 8
          2.1.3   Channel 12
    2.2   Modulation Techniques 13
    2.3   Noise Characteristic 14

Chapter 3   Positioning Algorithms and Systems 17

    3.1   Positioning Algorithms 17
          3.1.1   Triangulation 17
          3.1.2   Lateration 18
                3.1.2.1   Time-based Distance Estimation 18
                3.1.2.2   Distance Estimation Based on Signal
                          Attenuation 20
          3.1.3   Angulation 20
          3.1.4   Scene Analysis 21
          3.1.5   Proximity 22
    3.2   Indoor Positioning Systems 22
          3.2.1   Assisted-GPS 22
          3.2.2   Bluetooth 23
          3.2.3   Radio Frequency Identification 23
          3.2.4   Geo-Magnetism 24
          3.2.5   Ultra-Wideband 24
          3.2.6   Wireless Local Area Network 25
          3.2.7   Inertial Navigation System 25
          3.2.8   Zigbee 26
          3.2.9   Visible Light Communication 26

Chapter 4   Visible Light Positioning Systems 31

    4.1   System Configuration and Channel Access Method 31
          4.1.1   Time Division Multiple Access 31
          4.1.2   Basic Framed Slotted ALOHA 31
    4.2   Positioning Algorithm for 2-D Scenario 34

4.3   Linear Least Square Estimation...35
4.4   Positioning Algorithm for a 3-D Scenario...........................39

Chapter 5   Filtering Techniques in Positioning.....................................45

5.1   Kalman Filter..45
        5.1.1   Derivation of Kalman Filter.....................................45
        5.1.2   Kalman Filter in 2-D and 3-D Scenarios 47
5.2   Particle Filter 49
        5.2.1   Principle of Particle Filter......................................49
        5.2.2   Particle Filter in 2-D and 3-D Scenarios.................50
5.3   Gaussian Mixture Sigma-Point Particle Filter.....................51
        5.3.1   Principle 52
        5.3.2   GM-SPPF in 2-D and 3-D Scenarios........................54
5.4   Positioning Performance and Discussions............................56

Chapter 6   Three-Dimensional Positioning Based on Nonlinear Estimation
                   and Multiple Receivers 61

6.1   3-D Positioning Based on Nonlinear Estimation..................61
        6.1.1   Trust Region Reflective Algorithm 61
        6.1.2   3-D Positioning Algorithm 62
        6.1.3   Simulation and Results 64
6.2   3-D Positioning Based on Multiple Receivers......................64
        6.2.1   3-D Positioning Algorithms.....................................65
                  6.2.1.1   Received Signal Strength Method 65
                  6.2.1.2   Angle of Arrival Method 69
                  6.2.1.3   Combined RSS and AoA Using
                             Multiple Receivers....................................70
        6.2.2   Simulation and Results 74

Chapter 7   Impact of Multipath Reflections...79

7.1   Impulse Response 79
7.2   Deterministic Approach..81
7.3   MMC Approach...82
7.4   CDMMC Method..83
7.5   Analysis of Impulse Response..84
7.6   Power Intensity Distribution Analysis.................................89
7.7   Positioning Accuracy..90
7.8   Calibration Approaches 93
        7.8.1   Nonlinear Estimation 93
        7.8.2   Selection of LED Signals...98
        7.8.3   Decreasing the Distance Between LED Bulbs.......100

Chapter 8   OFDM Based Positioning Algorithm 105

      8.1   OFDM in Communication 105
      8.2   OFDM for VLC 106
            8.2.1   PAM-DMT 107
            8.2.2   DCO-OFDM 108
            8.2.3   ACO-OFDM 109
      8.3   OFDM in VLC positioning.. 110
            8.3.1   System Model 111
            8.3.2   Positioning Algorithm 114
      8.4   Simulation and Analysis 115
            8.4.1   Performance Comparison of Single- and
                     Multicarrier Modulation Schemes 116
            8.4.2   Effect of Signal Power on the Positioning
                     Accuracy 121
            8.4.3   Effect of Modulation Order on the Positioning
                     Accuracy 123
            8.4.4   Effect of Number of Subcarriers on the
                     Positioning Accuracy 123

Chapter 9   Sensor Fusion 129

      9.1   Methods and Goals 130
      9.2   Inertial Navigation System 131
            9.2.1   Stable Platform Systems 131
            9.2.2   Strap-down Systems... 131
      9.3   Inertial Sensors 133
            9.3.1   MEMS Accelerometer 134
            9.3.2   MEMS Gyroscope 135
      9.4   Realization by Kalman Filter.. 135
      9.5   Simulation and Results 137

Glossary 143

Index.. 147

Preface

This book provides a detailed overview of location-based optical wireless positioning and navigation services. The basic pedagogic methodology is to include fully detailed derivations from the basic principles. The text is intended to provide enough principles to guide the novice student, while having plenty of detailed materials to satisfy graduate students inclined to pursue research in the area. The book is intended to stress the principles of optical wireless positioning and navigation that are useful for a wide array of applications that these techniques are devised for. It is intended to serve as a possible textbook and reference for graduate students and a reference for practicing engineers.

ORGANIZATION OF THE BOOK

In Chapter 1, we offer an introduction to global positioning system (GPS) and its applications inside buildings. Due to difficulty of GPS signal penetration of building materials, radio frequency (RF) and optical wireless alternatives are considered.

This covers a broad array of issues like potential of solving complicated communications problems considering a shortage of radio frequency spectrum suitable for mobile applications, and suggestion of moving some of the less mobile applications to the optical frequencies range of infrared and visible light, RF interference, and necessity of transmission at high data rates, etc. by optical wireless systems. Optical wireless links can establish communications channels even millions of miles apart, as evidenced by the usage of optical links in space exploratory missions. For shorter terrestrial distances, optical wireless links in outdoor free space are a good choice for establishing pointed links a few miles apart.

In general, the optical wireless communications area of research did not receive much attention for several years, except in some military applications for the security that it offers. Yet the largest number of wireless devices ever sold, namely TV remote controls, use wireless infrared light in order to function. The advantages of using optical radiation over RF include:

- Virtually unlimited bandwidth with over 540 THz for wavelengths in the range of 200 nm–1550 nm. This band is unregulated and available for immediate utilization.
- Use of baseband digital technology.
- A small receiver (photo-detector) area provides spatial diversity that eliminates multipath fading in intensity modulation with direct detection links. Multipath fading degrades the performance of an unprotected RF link.
- Light is absorbed by dark surfaces, diffusely reflected by light-colored objects and directionally reflected from shiny surfaces. It does not penetrate opaque objects. This provides spatial confinement that prevents interference between adjacent cells operating in environments separated by opaque dividers.

- Spatial confinement of optical signals allows for secure data exchange without the fear of an external intruder listening in. This provides physical-layer security which is the safest type.
- No electromagnetic interference with other devices, making it very suitable for environments employing interference sensitive devices, such as hospitals, airports and factories, power plants, military and national security buildings.

Visible light (VL) applications are emerging technology areas that utilize the high-speed switching properties of VL light emitting diodes (LEDs) for wireless data applications with data rates higher than conventional local wireless networks and have additional benefits of:

- Sustainable solution for the current spectrum crunch.
- Energy efficiency in luminous efficacy – LEDs are far more efficient than incandescent and far more flexible than compact fluorescent lights (CFLs).

As LEDs increasingly displace incandescent lighting over the next few years, general applications of VL technology are expected to include wireless Internet access, vehicle-to-vehicle communications, broadcast from LED signage, machine-to-machine communications, positioning systems, and navigation. Furthermore, since smart LEDs have Internet protocol (IP) addresses, each will add a node to the Internet. Hence, "the most compelling story of how Internet of Light will transform our world is the one still being written: the future of lighting/communications/sensing/navigation and the birth of a new enterprise lighting network."[1]

As the next step to delve further into optical wireless positioning systems, we discuss some fundamentals of this technology in Chapter 2. We point out the differences between radio frequency-based and optical wireless communication systems and provide some details on optical transmitters and receivers.

In Chapter 3, we focus on positioning algorithms and systems. Considering the complex indoor environment, it is very difficult to model the signal propagation when severe multipath reflections exist and line-of-sight (LoS) condition is not satisfied. Indoor positioning algorithms that have been so far proposed can be classified into three groups: triangulation, scene analysis and proximity method. In this chapter, these algorithms along with their application on the short-range indoor localization will be discussed in detail. Positioning systems based on different techniques will be later introduced. Finally, an indoor positioning system based on visible light communication (VLC) as well as some recent research results is presented.

Our goal in Chapter 4 is to investigate the 2-D and 3-D positioning systems based on VLC. Two different channel access methods along with 2-D and 3-D positioning algorithms will be discussed. Finally, we evaluate the accuracy of the positioning systems in 2-D and 3-D scenarios.

Chapter 5 consists of the concept of filtering techniques that are used to further improve the positioning accuracy performance. In the positioning system, outliers

---

[1] M. Kavehrad, S. Chowdhury and Z. Zhou, "Short-Range Optical Wireless Theory and Applications," John Wiley & Sons Inc., December 2015.

may appear considering the basic framed slotted ALOHA (BFSA) failure case mentioned in Chapter 4. These large deviations heavily affect the positioning performance for several sequential estimates. Filtering techniques can mitigate these effects. In this chapter, we review three filtering techniques utilized in VLC positioning systems, namely Kalman filter, Particle filter, and Gaussian mixture sigma-point particle filter (GM-SPPF).

In Chapter 5, the horizontal positioning performance is improved by employing filtering techniques. However, the vertical positioning performance is not noticeably improved. In Chapter 6, we introduce algorithms based on nonlinear estimation and multiple receivers to improve the positioning performance particularly in the vertical direction.

In the previous chapters, the investigations are based on an LoS link, which is not very practical, taken the complex indoor environment into account. Multipath propagation phenomenon exists in the indoor wireless communications, i.e., the signals reach the receiver through more than one path. Multipath reflections degrade the communication quality as well as positioning accuracy. In Chapter 7, we will investigate the impact of multipath reflections on the positioning accuracy.

Orthogonal frequency-division multiplexing (OFDM) has been applied to indoor wireless optical communications in order to mitigate the effect of multipath distortion of the optical channel as well as increasing data rate. In Chapter 8, an OFDM VLC system is introduced which can be utilized for both communications and indoor positioning. We will demonstrate that the OFDM positioning system outperforms by 74% its conventional single carrier modulation scheme counterpart.

In Chapter 9, we discuss sensor fusion that addresses combining of sensory data or data derived from disparate sources such that the resulting information has less uncertainty than would be possible when these sources are used, individually. The term "uncertainty reduction" in this case can mean more accurate or more complete. We then describe an inertial navigation system (INS) as a navigation assistant that uses an inertial measurement unit (IMU), typically composed of accelerometers and gyroscopes, to continuously calculate the position, orientation, velocity and trajectory of a moving object. The INS available in smart phones and wearable devices frequently suffer from large positional errors in trajectories of motions and long-range displacements due to less-accurate IMU outputs—especially gyroscopes. Innovative techniques could improve the performance of INS with the advancement of IMU units by new physical methods, unique materials, and innovative fabrication techniques and/or with (i) the addition of different sensors, (ii) the sensor data fusion of external sensing and/or detection technologies, or (iii) smart computational algorithms for specific applications.

ACKNOWLEDGMENTS

I am grateful to all my doctoral students, in particular my co-author, Dr. Reza Aminikashani, who has contributed to this book through his thesis research.

Mohsen Kavehrad, CRKC LLC, Allentown, PA

1 Introduction and Overview

Location based services (LBSs) make use of the position information from mobile devices to provide related services [1]. The global positioning system (GPS) has been evolving since the 1970s, where the precise location information is made available through satellite infrastructures. Since then, LBS has been proposed and is being developed at a fast pace. From the late 1990s, techniques to realize LBS began to increase in popularity when mobile network services entered into a prosperous era.

With the location data of a user, many related services can be provided to make a user's life more convenient. For example, a user can easily get information about social events happening in a city where the user resides. For example, the nearest businesses or services, such as banks, restaurants, grocery stores and shopping malls can be located; turn-by-turn navigation is offered, and people are able to avoid traffic jams by personalizing their own routes considering cost, time consumption and distance. With the information on nearby friends, social networks are easily formed. Location based mobile advertisements can be promoted that benefit both businesses and customers. LBS can even assist health care for disabled people. Numerous mobile apps have been developed for LBS, such as yelp [2], Groupon [3], Foursquare [4] and Family Locator [5].

The number of large-scale buildings such as shopping malls, airports, museums and exhibition halls has been rapidly increasing due to the growth in urban populations. It is difficult for people to find their own locations, directions and friends in such areas. We all have had a hard time locating a new cafe inside in a bustling shopping center where we were supposed to meet our friends, or we have wandered the hallways of some office building trying desperately to find the meeting room we are expected at. Therefore, indoor localization systems can be lifesavers in countless circumstances. Furthermore, online location based advertisements, coupons and discount information make shopping easier. Fig. 1.1 shows an example of LBS in a shopping mall. In order to realize LBS, an economical and robust indoor positioning system is a key element that provides location information of users.

Indoor localization also finds its significant application in manufacturing to facilitate using robots. As artificial intelligence (AI) has been an emerging technique in recent years, using robots to replace human beings is an interest shared by some industries. In this way, production is increased, information of producing process is secured and some operations that are harmful for human involvement can be accomplished safely. In order to obtain the location of a robot, a precise localization system is indispensable [6, 7].

Although for outdoor environments, GPS can provide location information [8], it cannot be appropriately employed for indoor environments for the following reasons:

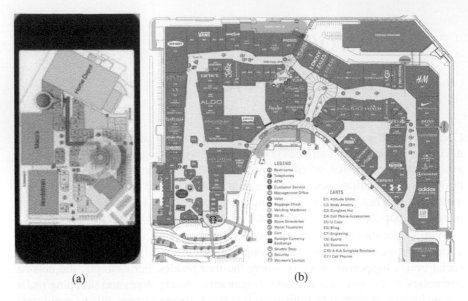

(a)   (b)

Figure 1.1: Examples of LBS: (a) Indoor navigation on the mobile phone. (b) Indoor map for user guidance.

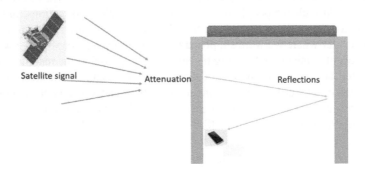

Figure 1.2: Attenuation of satellite signal.

First, as shown in Fig. 1.2, satellite signals are attenuated heavily when penetrating through solid walls, so a GPS signal is not detectable in many indoor locations. Second, GPS estimates the user coordinates from the travelling time information of the signal. For indoor environments, signals suffer severely from multipath reflections where the accuracy of time information is largely affected and the positioning performance is heavily degraded. Third, even for outdoor environments, the positioning accuracy of GPS is 7.8m of 95% confidence interval error and 4m of root mean square (RMS) error, which is not accurate enough for the application of indoor scenarios.

With the increasing demand of indoor location systems, several techniques have been taken into consideration, such as wireless local area network (WLAN), radio frequency identification (RFID), Zigbee, Bluetooth, assisted-GPS and visible light communication (VLC) [9, 10]. We will focus on VLC-based indoor localization, and introduce positioning algorithm and system configuration. Moreover, filtering techniques and a nonlinear estimation algorithm as well as orthogonal frequency-division multiplexing (OFDM) will be presented in order to improve the positioning accuracy and design robustness.

REFERENCES

1. Jochen Schiller and Agnès Voisard. *Location-based services*. Elsevier, 2004.
2. Yelp. https://www.yelp.com/.
3. Groupon. https://www.groupon.com/.
4. Foursquare. https://www.foursquare.com.
5. Life 360. https://www.life360.com/family-locator/.
6. Yu Zhou, Wenfei Liu, and Peisen Huang. Laser-activated RFID-based indoor localization system for mobile robots. In *Proceedings 2007 IEEE International Conference on Robotics and Automation*, pages 4600–4605. IEEE, 2007.
7. Bastian Bischoff, Duy Nguyen-Tuong, Felix Streichert, Marlon Ewert, and Alois Knoll. Fusing vision and odometry for accurate indoor robot localization. In *2012 12th International Conference on Control Automation Robotics & Vision (ICARCV)*, pages 347–352. IEEE, 2012.
8. Jay Farrell and Matthew Barth. *The global positioning system and inertial navigation*, volume 61. McGraw-Hill New York, 1999.
9. Hui Liu, Houshang Darabi, Pat Banerjee, and Jing Liu. Survey of wireless indoor positioning techniques and systems. *IEEE Transactions on Systems, Man, and Cybernetics, Part C (Applications and Reviews)*, 37(6):1067–1080, 2007.
10. Wenjun Gu, Weizhi Zhang, Mohsen Kavehrad, and Lihui Feng. Three-dimensional light positioning algorithm with filtering techniques for indoor environments. *Optical Engineering*, 53(10):107107, 2014.

2 Fundamentals of Visible Light Communication

Visible light communication (VLC) is an important application of optical wireless communications techniques and refers to unguided optical transmission via the use of light emitting diodes (LEDs). It has been gaining increasing attention in recent years as it is appealing for a wide range of applications such as indoor positioning. Indoor VLC is characterized by short transmission range and free from major outdoor environmental degradations such as rain, snow, building sway, and atmospheric turbulence [1].

As shown in Fig. 2.1, visible light falls in the range of the electromagnetic spectrum between infrared (IR) and ultraviolet (UV). The frequency band of VLC lies between 400 and 800 THz corresponding to wavelengths from about 390 nm to 700 nm.

2.1 SYSTEM MODEL

In this section, fundamental parts of VLC systems including the transmitter, the receiver and channel characteristics will be briefly discussed.

2.1.1 TRANSMITTER

As mentioned earlier, VLC systems mostly use semiconductor LEDs as transmitters due to their advantages and rapid developments in LED fabrication. Currently, lighting LEDs offer a transmission bandwidth range between 20 MHz to over 100 MHz, and as shown in Fig. 2.2, one LED bulb is composed of a number of LED chips. LED

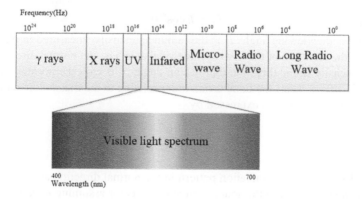

Figure 2.1: Electromagnetic spectrum.

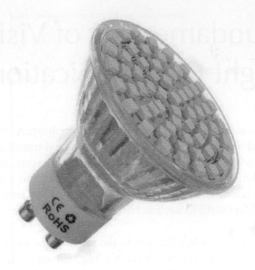

Figure 2.2: A sample LED bulb used as the transmitter in VLC.

lighting offers many advantages over conventional lighting, such as low power consumption, low cost, high luminance efficiency, long lifetime, etc. For these reasons, LEDs have become strong in the lighting market compared to traditional lighting solutions [2]. Besides these advantages, the intensity of the light emitted by an LED light can be modulated rapidly as it is a semiconductor device. This way, the LED combines its function of lighting with that of a data communication transmission system. It should be noted that LEDs are an incoherent light source, that is, LED light is not monochromatic and LED generates light with a broad bandwidth.

Fig. 2.3 shows the standard eye response curve, the warm white and ultra-white LED spectrum. The emission from an LED can be modeled using a generalized Lambertian radiant intensity given by [1]

$$I_\theta = I_0 \cos^m \theta \tag{2.1}$$

where θ is the viewing angle, I_0 is the radiance intensity (measured in photons/(s cm^2 sr)) in the direction of the LED symmetrical axis, i.e., the normal direction, and I_θ is the intensity in the direction that has an angle of θ with the normal direction. In Eq. (2.1), m represents the Lambertian order relating to the transmitter semi-angle $\varphi_{1/2}$ at half power, which is expressed by

$$m = -\frac{\ln 2}{\ln \left(\cos \varphi_{1/2} \right)} \tag{2.2}$$

Fig. 2.4 shows an LED radiation pattern in the normal direction and in the direction of an angle of θ from the source orientation vector assuming m equals one. In Fig. 2.4, $d\Omega$ is the solid angle encompassed by the receiver active area, and dA is the

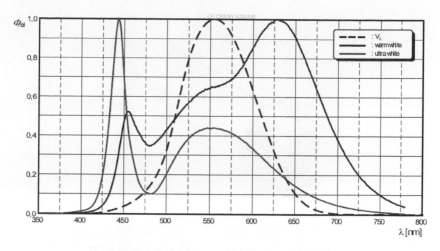

Figure 2.3: Relative spectral emission of an LED (LE CWUW S2W from OSRAM).

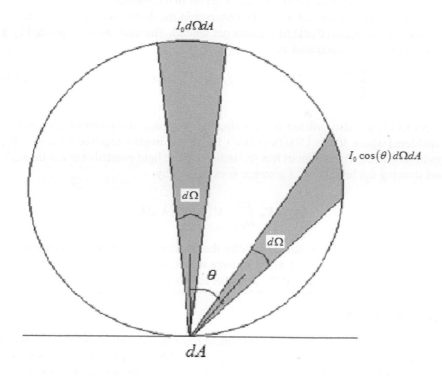

Figure 2.4: Emission rate (photons/s) in a normal and an angle of θ direction.

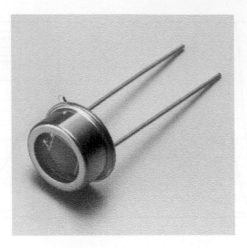

Figure 2.5: A sample PD used as the receiver in VLC.

emitting elementary area. Therefore, the observer in the normal direction will receive $I_0 d\Omega dA$ photons per second, while the observer in the direction of angle θ from the normal receives $I_0 \cos(\theta) d\Omega dA$ photons per second. The total intensity emitted by a source can be then calculated as

$$I = \iint_{source} I_0 \cos(\theta) d\Omega dA \qquad (2.3)$$

As LEDs are also utilized for illumination purposes, the luminance should be considered where 300 to 1500 lx (1 lx= 1 lm/m^2) is usually required for a working environment [3]. The luminous flux Φ_V indicating the light emitted over a solid angle and showing the brightness of a source is expressed by

$$\Phi_V = K_m \int_{\lambda_{\min}}^{\lambda_{\max}} V(\lambda) \Phi_e(\lambda) d\lambda \qquad (2.4)$$

In Eq. (2.4), K_m is the maximum visibility that is around 683 lm/W at 555 nm wavelength, $[\lambda_{\min}, \lambda_{\max}]$ is the wavelength range of visible light spectrum, $V(\lambda)$ is the standard luminosity curve, and $\Phi_e(\lambda)$ is the luminous flux per wavelength.

2.1.2 RECEIVER

A typical VLC receiver consists of an optical filter, amplification circuit and optical concentrators. A photodiode (PD) shown in Fig. 2.5 is commonly used to detect and convert the light to photo current which is proportional to the received optical power on the detection area. PD also has a limited bandwidth which ranges from several MHz to several GHz in a reverse proportion to the size of its active area. The larger the bandwidth is, the smaller the detection area is [4]. One of the most important

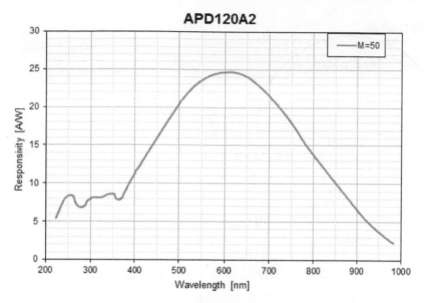

Figure 2.6: Detector responsivity of APD120A2.

parameters in PD performance is its responsivity which is expressed as

$$R = \frac{\eta q}{h f} \tag{2.5}$$

where η is the quantum efficiency of PD, q is the electron charge, h is Planck's constant, and f is the optical frequency. Fig. 2.6 demonstrates the responsivity curve of a PD (APD 120A2) manufactured by Thorlabs, Inc. [5]. For this PD, the responsivity reaches its highest value of 25 A/W at the wavelength of 600 nm and decreases rapidly when the wavelength becomes smaller or larger. In order to achieve a better communication performance, a proper PD should be selected with flat responsivity curve over the wavelengths generated by the LED.

In many indoor environments, there exists a strong ambient radiation arising from sunlight and non-transmitting lighting sources, which induces noise in a VLC receiver. The effects of ambient radiation and path loss can be mitigated by design of receivers having narrow optical bandwidth and large effective collection area. In order to maximize signal-to-noise ratio (SNR), it is therefore desirable to employ an optical concentrator to increase the effective collection area and optical filters to attenuate ambient radiation [6, 7].

Hemispherical lens is an important non-imaging concentrator. When longpass filtering is utilized, a planar filter is placed between the PD and the hemisphere as shown in Fig. 2.7a. However, when bandpass filtering is desired, the planer filter is not employed since θ, the angle at which light strikes the filter will be shifted by shifting ψ, the angle from which rays are received. This shift results in shifting the filter passband decreasing the effective signal-collection area as the results of a

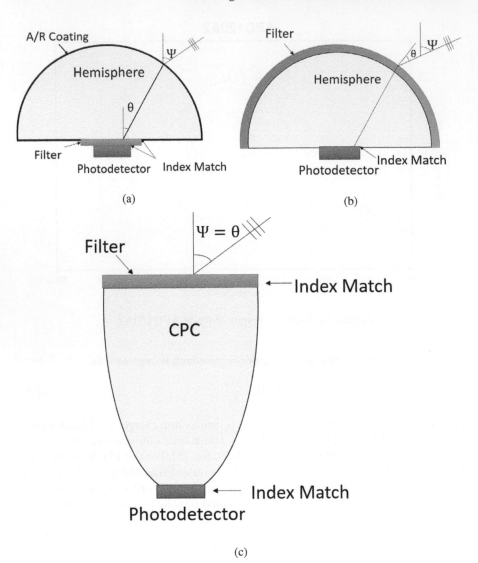

Figure 2.7: Optical concentrators: (a) hemisphere with planar optical filter, (b) hemisphere with hemispherical optical filter, and (c) CPC with planar optical filter.

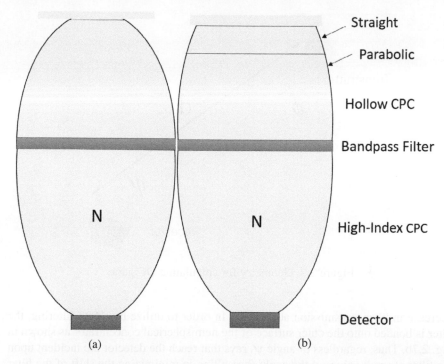

Figure 2.8: Combining CPCs with bandpass optical filters to improve acceptance angles. (a) Dielectric CPC combined with parabolic hollow CPC, and (b) Dielectric CPC combined with hollow CPC having a straight section.

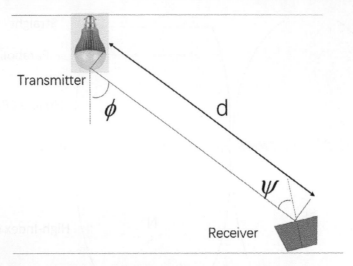

Figure 2.9: Geometry for calculating DC gain.

decrease in filter transmission at some ψ. In order to utilize bandpass filtering, the filter is bonded onto the outer surface of the hemispherical concentrator as shown in Fig. 2.7b. Thus, regardless of angle ψ, rays that reach the detector are incident upon the filter at small values of the angle θ resulting in minimizing the shift of the filter passband and maximizing its transmission.

The compound parabolic concentrator (CPC) is also widely used in VLC. It can provide a higher gain than the hemisphere concentrator but at the expense of a narrower field-of-view (FOV). As shown in Fig. 2.7c, a longpass or bandpass filter is placed on the front surface of the CPC. To achieve a wider FOV, a second inverted CPC can be placed on top of a traditional CPC as shown in Fig. 2.8a. In this way, radiations from an angle of 90° can be accepted by the upper CPC and then transferred to the angle within the FOV of the lower CPC. Fig. 2.8b shows further modification to CPC where a straight part is added on the upper CPC. Therefore, any angle between the FOV of the lower FOV and 90°can be received.

2.1.3 CHANNEL

Intensity modulation (IM) is commonly used in VLC where the intensity of the emitted light signal is modulated. Therefore, the modulated signal must be real and positive in the baseband. PD utilizes direct detection (DD) to collect the optical signal where the current is generated proportional to the light intensity striking on the effective receiving area. The frequency responses of VLC channels are relatively flat near direct current (DC). Therefore, for most purposes including indoor VLC localization, the single most important quantity characterizing a channel is the DC gain [7]. For a common link configuration shown in Fig. 2.9, the channel DC gain can be calculated as

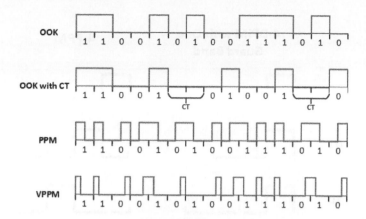

Figure 2.10: Two level modulation schemes.

$$H(0) = \begin{cases} \frac{m+1}{2\pi d^2} A \cos^m(\phi) T_s(\psi) g(\psi) \cos(\psi), & 0 \leq \psi \leq \Psi_c \\ 0, & \psi > \Psi_c \end{cases} \qquad (2.6)$$

where A is the receiving area of the PD, ϕ is the irradiance angle of the transmitter with respect to its normal axis, $T_s(\psi)$ is the signal transmission of optical filter and Ψ_c is the semi-angle of FOV. In Eq. (2.6), $g(\psi)$ is the concentrator gain given by

$$g(\psi) = \begin{cases} \frac{n_c^2}{\sin^2(\psi_c)}, & 0 < \psi < \Psi_c \\ 0, & \psi > \Psi_c \end{cases} \qquad (2.7)$$

where n_c is the concentrator refractive index. The received optical signal power P_r is proportional to the average transmitted power P_t as

$$P_r = H(0) P_t \qquad (2.8)$$

2.2 MODULATION TECHNIQUES

IM/DD communication systems use intensity modulation techniques such as on-off keying (OOK) or pulse position modulation (PPM). In OOK, the optical transmitter is on during the whole bit interval when 1 is transmitted, and is off when 0 is transmitted. On the other hand, in binary pulse position modulation (BPPM), the optical transmitter is on during a half of the BPPM bit interval (i.e., signal slot) and is off during the other half (i.e., non-signal slot) [8]. To achieve the desired dimming requirement, OOK with dimming by the insertion of compensation time (CT) and variable pulse position modulation (VPPM) are used that optimize data and bit error rates without sacrificing illumination quality [9]. Particularly, in OOK with CT, an information subframe is followed by a CT subframe, which adjusts the "on/off"

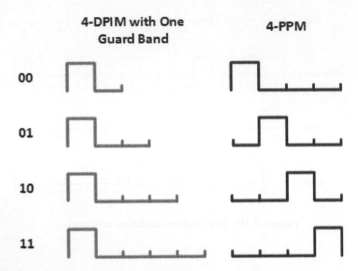

Figure 2.11: DPIM with one guard band and 4-PPM modulation schemes.

duration to reach the target dimming percentage. VPPM is a modified PPM scheme, which increases or decreases the pulse width considering the required dimming level. Fig. 2.10 further demonstrates these modulation schemes.

Digital pulse interval modulation (DPIM) is another alternative modulation scheme to PPM where the symbol length is variable and an additional guard slot is usually added to each symbol to avoid the adjacent zeroes as shown in Fig. 2.11 [10]. There are two advantages of DPIM: first, the power efficiency or the bandwidth efficiency is high; second, there is no synchronization requirement.

2.3 NOISE CHARACTERISTIC

It is well understood that the dominant noise sources in indoor optical wireless communications systems are the background light induced shot noise and thermal noise [7]. The sum of contributions from shot noise and thermal noise follows a Gaussian distribution with the variance of

$$N = \sigma_{shot}^2 + \sigma_{thermal}^2 \tag{2.9}$$

where $\sigma_{thermal}^2$ and σ_{shot}^2 denote the variance of thermal and shot noise, respectively. The shot noise variance is obtained by

$$\sigma_{shot}^2 = 2q\gamma(P_{rec})B + 2qI_{bg}I_2 \tag{2.10}$$

where B is equivalent noise bandwidth, I_{bg} is background current, I_2 is noise bandwidth factor, γ is detector responsivity, and P_{Rec} denotes all the received power including the ambient light such as sunlight or other lighting fixtures. It is assumed

Table 2.1
Numerical values for noise parameters

Parameters	Value
Electronic charge (q)	1.602×10^{-19} C
Background current (I_{bg})	5100 μ A
Noise bandwidth factor (I_2)	0.562
PD responsivity (γ)	0.54 A/W
Boltzmann's constant (k)	1.381×10^{-23} K^{-1}
Absolute temperature (T_K)	295 K
Open-loop voltage gain (G_o)	10
Fixed capacitance per unit area (η)	112 pf/cm^2
Noise factor of FET channel (Γ)	1.5
FET transconductance (g_m)	30 mS
Noise bandwidth factor (I_3)	0.0868

here that a p-i-n/field-effect transistor (FET) transimpedance receiver is used and the noise contributions from gate leakage current and $1/f$ noise are negligible [11].

The thermal noise variance is given by

$$\sigma^2_{thermal} = \frac{8\pi k T_K}{G_o}\eta A I_2 B^2 + \frac{16\pi^2 k T_K \Gamma}{g_m}\eta^2 A^2 I_3 B^3 \qquad (2.11)$$

where the two terms indicate feedback-resistor noise, and FET channel noise, respectively. In Eq. (2.11), k is Boltzmann's constant, T_K is absolute temperature, G_o is the open-loop voltage gain, η is the fixed capacitance of photo detector per unit area, Γ is the FET channel noise factor, g_m is the FET transconductance, and I_3 is noise bandwidth factor.

In the numerical examples presented in this book, the values shown in Table 2.1 are used.

SUMMARY

This book focuses on indoor localization with VLC technology, which overcomes some problems that currently exist in radio frequency (RF)-based technology such as multipath reflections, shortage of radio frequency spectrum and electromagnetic radiation interference. In addition, the book presents state-of-the-art research on three main aspects, namely: (i) it constructs the concept and the model for the systems and the sub-systems; (ii) it covers positioning algorithms, as the main issue in indoor localization, and (iii) several remedies are proposed to further improve the positioning accuracy performance.

In this chapter, fundamental physical layer end-to-end transmission link for visible light communications systems are discussed. This includes a description of transmitter sub-systems, receiver elements and channel characteristics plus some candid modulation techniques traditionally adopted. In addition, receiver front-end noise characteristics are detailed.

REFERENCES

1. Mohsen Kavehrad, MI Sakib Chowdhury, and Zhou Zhou. *Short-Range Optical Wireless: Theory and Applications*. John Wiley & Sons, 2016.
2. Mohsen Kavehrad. Sustainable energy-efficient wireless applications using light. *IEEE Communications Magazine*, 48(12):66–73, 2010.
3. Toshihiko Komine and Masao Nakagawa. Fundamental analysis for visible-light communication system using led lights. *IEEE Transactions on Consumer Electronics*, 50(1):100–107, 2004.
4. Gerd Keiser. *Optical Communications Essentials (Telecommunications)*. McGraw-Hill Professional, 2003.
5. Thorlabs. http://www.thorlabs.com/thorcat/MTN/APD120A2-Manual.pdf.
6. Keang-Po Ho and Joseph M Kahn. Compound parabolic concentrators for narrowband wireless infrared receivers. *Optical Engineering*, 34(5):1385–1396, 1995.
7. Joseph M Kahn and John R Barry. Wireless infrared communications. *Proceedings of the IEEE*, 85(2):265–298, 1997.
8. Mohammadreza A Kashani, Majid Safari, and Murat Uysal. Optimal relay placement and diversity analysis of relay-assisted free-space optical communication systems. *Journal of Optical Communications and Networking*, 5(1):37–47, 2013.
9. John Gancarz, Hany Elgala, and Thomas DC Little. Impact of lighting requirements on vlc systems. *IEEE Communications Magazine*, 51(12):34–41, 2013.
10. Zabih Ghassemlooy, AR Hayes, NL Seed, and ED Kaluarachchi. Digital pulse interval modulation for optical communications. *IEEE Communications Magazine*, 36(12):95–99, 1998.
11. Andrew P Tang, Joseph M Kahn, and Keang-Po Ho. Wireless infrared communication links using multi-beam transmitters and imaging receivers. In *Proceedings of ICC/SUPERCOMM'96-International Conference on Communications*, volume 1, pages 180–186. IEEE, 1996.

3 Positioning Algorithms and Systems

Considering the complex indoor environment, it is very difficult to model the signal propagation when severe multipath reflections exist and a line-of-sight (LoS) condition is not satisfied. In this chapter, these algorithms along with their application on the short-range indoor localization will be discussed in detail. Positioning systems based on different techniques will be later introduced. Finally, an indoor positioning system based on visible light communication (VLC) as well as some recent research results is presented.

3.1 POSITIONING ALGORITHMS

Indoor positioning algorithms that have been so far proposed can be classified into three groups: triangulation, scene analysis and proximity method as shown in Fig. 3.1. In the following, we will describe these techniques in more detail.

3.1.1 TRIANGULATION

Triangulation generally refers to positioning techniques using the geometric properties of triangles for location estimation. In this method, coordinates of the target are estimated according to the geometric properties of triangles. Triangulation has been implemented by two algorithms: lateration and angulation.

Lateration methods use the distance of the target from multiple reference points to estimate its location, while angulation locates the target by analyzing the incident angles of the received signals relative to the reference points.

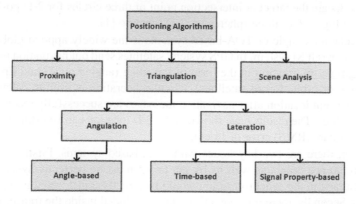

Figure 3.1: Indoor localization methods and algorithms.

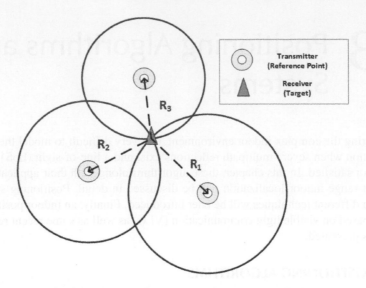

Figure 3.2: Positioning based on TOA/RTOF/RSS.

3.1.2 LATERATION

Lateration methods estimate the target location by measuring its distances from multiple reference points and utilize two types of measurements: time of arrival (TOA) and received signal strength (RSS).

3.1.2.1 Time-based Distance Estimation

The light speed is constant in the air. Therefore, the distance from the target to the reference points can be deducted from the travel time of the light signal. In time-based systems, TOA measurements with respect to at least three reference points are required to locate the target at intersection point of three circles for 2-D positioning as shown in Fig. 3.2, or three spheres for 3-D scenario [1].

An excellent example of TOA-based systems is the widely applied global positioning system (GPS) system. In GPS system, satellites send out navigation messages containing time information in the form of repeating ranging codes and Ephemeris (information of orbits for all satellites). Circular lateration is used to calculate the receiver's current location after navigation messages are successfully received from several satellites. The results have shown that a 2-D location can be estimated with a root mean square (RMS) error of 14 cm.

However, there are two key issues with TOA-based systems. Firstly, all clocks used by reference points as well as by the target must be perfectly synchronized. Any inaccuracy in the synchronization would be directly transformed into positioning errors. Secondly, there must be a time stamp included inside the transmitted signal potentially requiring extra cost in terms of data rate. To satisfy numerous indoor

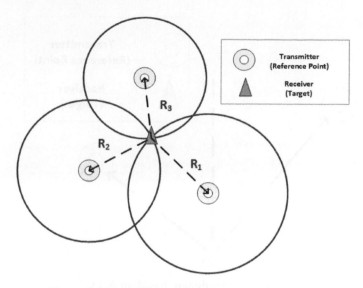

Figure 3.3: Positioning using hyperbolic lateration.

application demands, positioning accuracy range of meters to centimeters is required, which means all clocks in TOA-based systems need to be synchronized at least at the level of several nanoseconds. As a result, complexity and cost of such systems cannot be properly addressed and research on VLC positioning utilizing TOA measurements has been very limited. Simulation results in [2] give Cramer-Rao bound which indicates, depending on system settings, around $2 \sim 5$ cm positioning accuracy may be obtained considering only shot noise.

To address these issues, hyperbolic lateration methods usually utilizing time-difference-of-arrival (TDOA) measurements are used. Different from TOA, TDOA-based systems measure the difference in time at which signals from different reference points arrive. These signals must be transmitted at the same time; therefore, all the transmitters as reference points have to be synchronized precisely. In VLC, this can be easily realized because light emitting diode (LED) bulbs are in close proximity. On the other hand, the receiver does not have to be synchronized with transmitters since it is not taking measurements of the absolute time of arrival. Moreover, no time stamp is required to be labeled in the transmitted signal.

Same as in TOA-based systems, three reference points are needed to perform 2-D or 3-D positioning. Since a single TDOA measurement provides a hyperboloid on a 2-D plane or a hyperbola in a 3-D space, two TDOA measurements from three reference points are needed to locate the target by finding the intersection point as shown in Fig. 3.3. A TDOA-based system using ultra-wideband (UWB) technology experimentally demonstrated a positioning accuracy of about 20 cm [3].

As a conclusion, for short range indoor positioning, time-based methods are expensive to deploy and easy to be affected by time delay, reflections and measurement precision.

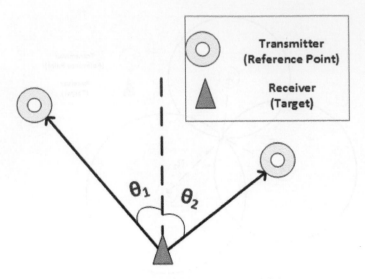

Figure 3.4: Positioning based on AOA.

3.1.2.2 Distance Estimation Based on Signal Attenuation

For positioning systems with the time-based distance estimation, the hardware complexity and cost is high and these systems can only be used in some specific areas. RSS information is therefore more commonly used in lateration techniques. These systems measure the received signal strength, and calculate the propagation loss that the emitted signal has experienced. Range estimation is then made by employing a path loss model. Available RSS-based methods using WLAN technology can provide accuracy of 4 m with 90% confidence [4]. VLC-based systems provide better positioning accuracies due to weaker multipath effects compared to radio-wave approaches, leading to a better estimation of the received signal strength.

3.1.3 ANGULATION

When a directional antenna or an array of antennas are used, the incident angles from several transmitters can be detected as the angle of arrival (AOA) information. Target is then located by finding intersection of direction lines. Theoretically, only two reference points are needed to perform 2-D and three for 3-D positioning. The most important feature of AOA-based systems is that no synchronization is needed between reference points and target. As shown in Fig. 3.4, utilizing the detected incident angles θ_1 and θ_2, the target location is estimated at the intersection point.

In radio frequency (RF) domain, AOA technique is mainly applied in UWB systems and in [5,6] authors propose a joint TOA and AOA estimation algorithm yielding 10 cm to 35 cm positioning accuracy, depending on the type of pulse employed.

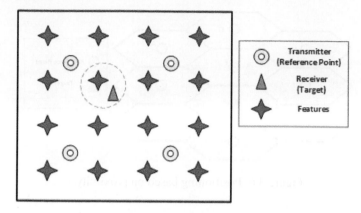

Figure 3.5: Positioning based on scene analysis.

In [7], the AOA information is detected with three antenna arrays for a radio frequency identification (RFID) based system.

Imaging receivers such as front-facing cameras on smart phones can be easily used to detect the AOA of the incoming optical signal. However, large field-of-view (FOV) cameras or new lighting infrastructures are required to achieve satisfying performance. The positioning accuracy is also degraded when the target moves away from reference points due to the limited resolution of imaging receiver. To address this issue, fly-eye receivers can be utilized [8, 9].

Another shortcoming of AOA methods is the high complexity and cost of imaging receivers, though their size is much smaller than antenna arrays used in radio-wave approaches. In [10], the authors have showed that with an imaging receiver with a resolution of 1296 × 964 pixels, the positioning accuracy of 5 cm can be reached. Furthermore, in [11], researchers have taken multipath reflections into consideration and proposed a two-phase positioning algorithm which utilizes both RSS and AOA information. It has been shown using computer simulations that a median accuracy of 13.95 cm can be achieved.

3.1.4   SCENE ANALYSIS

The scene analysis approach has been usually used in the RF-based systems. As shown in Fig. 3.5, it includes two stages in which first a site survey is conducted to collect features associated with every position in the scene as fingerprints. Target location is then identified by matching real-time measurements to these features. Factors that can be used as fingerprints include but are not limited to all measurements mentioned earlier, i.e., TOA, TDOA, RSS and AOA. Because of complexity and other concerns, the RSS information from the nearby transmitters are usually considered as fingerprints in both RF and optical domain.

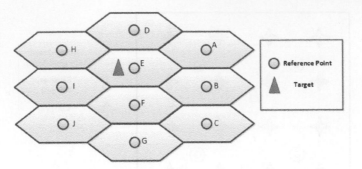

Figure 3.6: Positioning based on proximity.

3.1.5 PROXIMITY

Proximity is a straightforward approach to estimate the location based on a dense grid of reference points with known coordinates. As shown in Fig. 3.6, the target is considered to share the same coordinates with reference point A. When a mobile target detects a signal from a single reference point, it is considered co-located with it. If more than one signal are detected, the receiver is considered to be collocated with the reference point sending the strongest signal.

Therefore, proximity systems using light sources as transmitters theoretically provide accuracy no more than the resolution of the grid itself. Notice that when dense grids are used, beam profiles of light sources must be optimized in order to minimize inter-source interference, which leads to extra positioning errors. The proximity algorithm is easy to be implemented in Bluetooth, Zigbee, and RFID systems. In [12,13], the authors proposed and experimentally demonstrated a proximity-based indoor positioning system. Apart from positioning provided by the visible light LED, a Zigbee wireless network is used to transmit the location information to the main node, as well as to extend the working range.

3.2 INDOOR POSITIONING SYSTEMS

Based on the above positioning algorithms, several systems have been proposed with different techniques. Current indoor wireless positioning systems mainly include two categories. The first category is to build up a network infrastructure which primarily focuses on localization purposes. In this category, relatively high accuracy can be achieved while the deployment cost is high. The second category utilizes the existing network to locate a target, which is usually economical but the accuracy is not satisfactory. In the following part, we provide a general overview of these systems.

3.2.1 ASSISTED-GPS

As GPS signal attenuates dramatically when passing through solid walls, there are mainly two approaches of assisted-GPS to make it applicable for indoor environments. GPS utilizes the time-based information to estimate the distances. In one approach, the sensitivity of the receiver is improved so that the GPS signals can be detected inside a building or even in a deep urban area. In [14], an assisted-GPS receiver is designed with reduced squaring loss so that the weak GPS signal whose strength is less than -150 dBm can be detected within 2~3 seconds. Another approach makes use of mobile station based services where a location server is employed. The satellite signals are simultaneously detected at the mobile station with a partial GPS receiver. Therefore, the weak GPS signals inside a building can be detected. In [15], a roof-antenna with known location is used to get access to satellite signals in the LoS condition. The user obtains the data from the antenna through wireless network.

3.2.2 BLUETOOTH

Bluetooth is a wireless technique which covers short distances and utilizes the industrial, scientific and medical (ISM) band from 2.4 GHz to 2.485 GHz. Bluetooth has been proposed for indoor localization for the following two reasons. First, Bluetooth is ubiquitously embedded in most phones, tablets and personal digital assistants (PDAs). Therefore, in the receiver side, no more hardware cost is introduced. Second, a Bluetooth signal usually ranges from 10 m to 15 m, which makes it an appropriate technique to employ proximity algorithms, as signals from different transmitters will not interfere too much with each other. Particularly, since Bluetooth 4.0 and Bluetooth low energy (BLE) have been developed in recent years, an indoor localization with low energy consumption can be achieved. Thus, the Bluetooth-based indoor localization has become a practical approach to locate users carrying Bluetooth embedded devices. In [16], authors have proposed a BLE-based localization scheme utilizing the collected received signal strength indication (RSSI) measurements to generate a small region in which the object is guaranteed to be found. It is different than most of the existing localization methods that attempt to find the specific location of the object under investigation.

In another system, a device inquiry scheme and service discovery protocol have been designed so that the connections can be robustly established [17]. The positioning algorithm is based on scene analysis where data is trained in the first phase and location is estimated in the second phase using the Weibull distribution model.

3.2.3 RADIO FREQUENCY IDENTIFICATION

RFID readers utilize the distribution of electromagnetic fields to identify and track tags which are attached to the users. Passive tags are small and inexpensive, which can operate without a battery, but their reading range is only $1 \sim 2$ m. Active RFID tags can actively transmit their identification (ID) up to tens of meters. A well-known RFID positioning system is LANDMARC, using active RFID tags operating at the frequency of 308 MHz. The LANDMARC approach requires signal strength information from each tag to readers, if it is within the detectable range. In [18], extra tags are used so that the readers' number can be decreased. In their method, signal

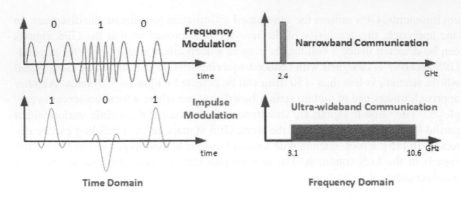

Figure 3.7: Time and frequency property of UWB.

strength information and K-nearest neighbors algorithm are used to locate the RFID tags. Recently, RFID localization system has been used in various markets such as retail, oil and gas, manufacturing, healthcare and event security and safety.

3.2.4 GEO-MAGNETISM

The rapid development of sensor techniques has been enabling the utilization of geo-magnetism in indoor localization. Theoretically speaking, the magnetic field of each point on the earth is unique although their differences are difficult to be detected with normal devices. The structural steel elements in a building disturb the magnetic fields and enhance these differences in indoor environments. As these magnetic fields are spatially varied but temporally stable, geo-magnetism is appropriate to be used in indoor localization. The magnetic field is usually measured with an array of e-compasses, and then matched to pre-acquired magnetic map. In [19], a positioning system has been proposed including a magnetic fingerprint map and a client device which detects the magnetic signature of the current position. The accuracy within 1 m in 88% of the time is experimentally demonstrated. In [20], Monte Carlo localization algorithm has been proposed along with a series of global localization experiments conducted in four arbitrarily selected buildings to demonstrate the feasibility of the geo-magnetism based system.

3.2.5 ULTRA-WIDEBAND

As shown in Fig. 3.7, the bandwidth of UWB in frequency domain is ultra-wide enabling the use of pulses with bandwidth of more than 500 MHz for impulse modulation. Since a wide range of frequency components is included, the probability that the wave transmits through or around obstacles is increased. Accordingly, the system reliability is improved. As the UWB signal usually spreads over a large frequency range, i.e., from 3.1 to 10.6 GHz, the power spectral density (PSD) is decreased and the interference is reduced compared to other RF-based systems. For UWB localization systems, time-based schemes namely TOA and TDOA provide good

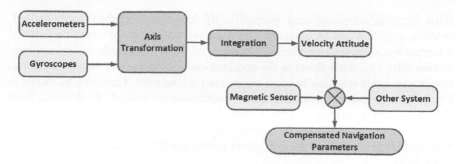

Figure 3.8: Block diagram of INS.

accuracy due to the high time resolution (large bandwidth) of UWB signals. In [21], UWB tags have been combined with transmitters while the UWB reference nodes are collocated with both transmitters and receivers. Impulse radio UWB transceivers are employed as test setup and TDOA information are obtained at distinct positions to locate the target.

3.2.6 WIRELESS LOCAL AREA NETWORK

As wireless local area network (WLAN) is a popular method for indoor communication; currently, localization based on WLAN is a hot research topic. As the WLAN signal is transmitted in the band of 2.4 GHz, the performance is easily affected by the orientation of antennas, obstacles, positions of reference points and multipath reflections. The accuracy of the typical WLAN positioning system ranges from 3 m to 30 m, and the sampling rate is up to a few seconds. A common applied algorithm in the WLAN-based positioning system is scene analysis. A site survey is conducted first to collect RSS fingerprints and then the target's location is found with the matching algorithms. Spectral clustered time-stamped RSS data are usually utilized to describe the layout of indoor environments [22]. Spectral clustering is performed to classify the pre-collected RSS data, and logic graphs are constructed and mapped into ground-truth graphs. In the online phase, area-level positioning can be achieved by estimating the smallest Euclidean distances between the current detected RSS information and the centers of the nodes.

3.2.7 INERTIAL NAVIGATION SYSTEM

Inertial navigation system (INS) is based on a straightforward principle which is integrating the velocity and the orientation at each time unit to obtain the current location. The platform module contains motion sensors so that the postures such as roll, yaw and pitch angle are detected. In inertial measurement unit (IMU), accelerometers are used to measure the linear acceleration, rotation sensors are applied to detect the angular velocity and magnetic sensors are utilized for calibration. INS techniques have a long history and originally were used on vehicles such as ships, aircrafts, submarines and spacecrafts. By dead reckoning the velocity and orientation, positions can be obtained with a known starting point where no external reference measurements are needed. However, all the INS methods suffer from integration drift, i.e., small errors in the acceleration and the angular velocity accumulate progressively resulting in larger errors after a long time. Therefore, as shown in Fig. 3.8, the estimated coordinates must be calibrated periodically by the input from other localization systems. In [23], a strap-down INS has been proposed containing an IMU which utilizes a dead reckoning algorithm. Bluetooth v4.0 technique has been employed to calibrate the drift errors so that the INS can deliver high accuracy positioning.

We shall return to this discussion in Chapter 9 and expand on the INS methodologies as a necessary part of data fusion.

3.2.8   ZIGBEE

Based on the IEEE 802.15.4 protocol, the Zigbee technique is able to set up a personal network area using small and low-power radios. A Zigbee system is composed of coordinators, routers and end devices. Zigbee is a simpler and less expensive technology than Bluetooth and Wi-Fi, and is widely used in traffic control systems, smart home display and wireless switches. A Zigbee network is easy to expand, and its transmission distance is from 10 m to 100 m making it appropriate to be used in indoor localization system. Since both WLAN and Bluetooth work under the 2.4 GHz frequency, RSS information of Zigbee signal can be easily interfered with, resulting in decreasing the positioning accuracy. Therefore, two localization mechanisms are proposed to address this shortcoming. The first one is based on the localization map where all the nodes, except the one collocated with the user, are presumed to have fixed known coordinates. A site survey is conducted to collect the RSS fingerprints as vector data and stored in a database. A mobile receiver estimates the current location by matching the current RSS to the fingerprint maps with K-nearest neighbors algorithm. Another mechanism is to estimate the distance information with Markov chain inference, where mobile target behavior is estimated. Optimal decision is computed for the uncertainty of the behavior, and the estimated distance is used to update the predefined positions [24].

3.2.9   VISIBLE LIGHT COMMUNICATION

RF-based techniques deliver positioning accuracy from tens of centimeters to several meters. However, this amount of accuracy is not sufficient for practical applications such as location detection of products inside large warehouses to automate some of the inventory management processes. Apart from the relatively poor accuracy of indoor positioning achievable by RF-based techniques, they also add to the electromagnetic (EM) interference. For these reasons, the techniques based on VLC are gaining more attraction [25].

VLC-based techniques employ fluorescent lamps and LEDs. Utilizing VLC, positioning services can be offered universally wherever the lighting infrastructures exist, even in some environments that RF radiation is dangerous or even forbidden, such as hospitals with a lot of medical equipment and nucleus industries. VLC-based localization systems can fit in these RF-prohibited areas perfectly since no electromagnetic interference is generated. Most of the VLC-based techniques use LEDs as the light source, since they can be modulated more easily compared to fluorescent lamps, and location data can be transmitted in a simpler way. Hence, proposed indoor positioning techniques based on VLC and LEDs are excellent options. Moreover, LEDs are currently being installed in most buildings, especially larger ones, e.g., museums and shopping malls, as the primary lighting source instead of fluorescent lamps since they have the advantage of much longer life and lower operating cost.

In addition, as mentioned in the previous sections, in the two categories of current positioning systems, the systems using the existing network cannot provide satisfactory positioning accuracy while the systems using their own infrastructures bring up the deployment cost. Indoor positioning systems based on a VLC technique have both advantages of these two categories. As it utilizes the existing illumination facilities, the deployment cost is minimized. At the same time, as light wave has relatively shorter wavelength than radio wave, it suffers less from multipath reflections resulting in a satisfactory positioning accuracy.

In the VLC positioning system, LEDs are the reference points and a photodiode (PD) collocated with the user is the target. In [2], a TOA-based VLC positioning system has been proposed where perfect synchronization between the transmitter and the receiver is assumed. The theoretical limit on the estimation accuracy has been analyzed by deriving the Cramer-Rao bound for the windowed sinusoidal waves. In [26], a TDOA-based VLC localization system using a coherent heterodyne detection has been proposed. Gaussian noise has been considered for the mathematical model where three stages of estimation algorithms have been used to minimize the noise impairments [26]. Moreover, RSS-based lateration systems have been widely investigated due to the simplicity of the distance estimation algorithm as well as low requirements of the detection devices [1]. Angulation technique has been also used to realize indoor localization in the VLC system, where an image sensor usually acts as the receiver to obtain AOA information. In order to distinguish the transmitters, the colored LEDs are used [10]. In a scene analysis based light positioning system, a new value called correction sum ratio is defined [27]. This value is obtained with a site survey. When a mobile terminal knows its current value at an arbitrary location, its coordinates can be detected with a positioning accuracy of around 1 cm. In other work, proximity techniques have been utilized in a small experimental area where four LEDs were used as the transmitters with different IDs [28].

In addition to the above methods, VLC can also be combined with other techniques to further improve the system performance. In [12], Zigbee technology has been combined with VLC so that long distance positioning can be achieved. In [29], by employing 6-axe sensors, i.e., geomagnetic sensor and gravity acceleration sensor, a switching estimated receiver positioning system has been proposed to achieve high accuracy. In [30], an accelerometer has been used to develop a three-dimensional

VLC based localization system without the knowledge of the receiver height. Furthermore, filtering techniques such as Kalman filter, particle filter and Gaussian mixture sigma-point particle filter (GM-SPPF) have been used to further increase the positioning accuracy and prevent large deviations [31, 32]. We will expand on the latter topics in the upcoming chapters.

SUMMARY

In the last chapter, the VLC transmission link was introduced in details. Since an LED acts as the transmitter, the radiation pattern was detailed. Receiver is usually a PD to detect the received signal strength and its FOV decides how many signals it can detect. If the channel is a LoS link, channel direct current (DC) gain is the main factor that determines the received power. Channel DC gain is related to the transmitter and receiver properties and the distance. The noise in the channel induces positioning errors that is mainly composed of shot noise and thermal noise following Gaussian distribution.

In the current chapter, we covered the basics of the positioning algorithms such as triangulation, scene analysis and proximity methods. The positive and negative aspects of these algorithms were analyzed. Different candid techniques used for indoor localization were described, such as Assisted-GPS, UWB, Zigbee, WLAN and Bluetooth. Finally, the advantages of the positioning system based on VLC were detailed.

REFERENCES

1. Wenjun Gu, Weizhi Zhang, Mohsen Kavehrad, and Lihui Feng. Three-dimensional light positioning algorithm with filtering techniques for indoor environments. *Optical Engineering*, 53(10):107107, 2014.
2. Thomas Q Wang, Y Ahmet Sekercioglu, Adrian Neild, and Jean Armstrong. Position accuracy of time-of-arrival based ranging using visible light with application in indoor localization systems. *Journal of Lightwave Technology*, 31(20):3302–3308, 2013.
3. Koichi Kitamura and Yukitoshi Sanada. Experimental examination of a UWB positioning system with high speed comparators. In *2007 IEEE International Conference on Ultra-Wideband*, pages 927–932. IEEE, 2007.
4. Yuhong Liu and Yaokuan Wang. A novel positioning method for WLAN based on propagation modeling. In *2010 IEEE International Conference on Progress in Informatics and Computing*, volume 1, pages 397–401. IEEE, 2010.
5. Lorenzo Taponecco, Antonio Alberto D'Amico, and Umberto Mengali. Joint toa and aoa estimation for uwb localization applications. *IEEE Transactions on Wireless Communications*, 10(7):2207–2217, 2011.
6. Naveed Salman, Muhammad Waqas Khan, and Andrew H Kemp. Enhanced hybrid positioning in wireless networks ii: AOA RSS. In *2014 International Conference on Telecommunications and Multimedia (TEMU)*, pages 92–97. IEEE, 2014.
7. Salah Azzouzi, Markus Cremer, Uwe Dettmar, Thomas Knie, and Rainer Kronberger. Improved AOA based localization of UHF RFID tags using spatial diversity. In *2011 IEEE International Conference on RFID-Technologies and Applications*, pages 174–180. IEEE, 2011.

8. G Yun and M Kavehrad. Indoor infrared wireless communications using spot diffusing and fly-eye receivers. *Canadian Journal of Electrical and Computer Engineering*, 18(4):151–157, 1993.

9. G Yun and M Kavehrad. Spot-diffusing and fly-eye receivers for indoor infrared wireless communications. In *1992 IEEE International Conference on Selected Topics in Wireless Communications*, pages 262–265. IEEE, 1992.

10. Toshiya Tanaka and Shinichro Haruyama. New position detection method using image sensor and visible light leds. In *2009 Second International Conference on Machine Vision*, pages 150–153. IEEE, 2009.

11. Gregary B Prince and Thomas DC Little. A two phase hybrid RSS/AOA algorithm for indoor device localization using visible light. In *2012 IEEE Global Communications Conference (GLOBECOM)*, pages 3347–3352. IEEE, 2012.

12. Yong Up Lee and Mohsen Kavehrad. Two hybrid positioning system design techniques with lighting LEDs and ad-hoc wireless network. *IEEE Transactions on Consumer Electronics*, 58(4):1176–1184, 2012.

13. YU Lee, S Baang, J Park, Z Zhou, and M Kavehrad. Hybrid positioning with lighting LEDs and Zigbee multihop wireless network. In *Broadband Access Communication Technologies VI*, volume 8282, page 82820L. International Society for Optics and Photonics, 2012.

14. Deok Won Lim, Sang Jeong Lee, and Deuk Jae Cho. Design of an assisted GPS receiver and its performance analysis. In *2007 IEEE International Symposium on Circuits and Systems*, pages 1742–1745. IEEE, 2007.

15. Al'Khateeb Anwar, Gragopoulos Ioannis, and Fotini-Niovi Pavlidou. Indoor location tracking using a GPS and Kalman filter. In *2009 6th Workshop on Positioning, Navigation and Communication*, pages 177–181. IEEE, 2009.

16. Yixin Wang, Qiang Ye, Jie Cheng, and Lei Wang. RSSI-based bluetooth indoor localization. In *2015 11th International Conference on Mobile Ad-hoc and Sensor Networks (MSN)*, pages 165–171. IEEE, 2015.

17. Ling Pei, Ruizhi Chen, Jingbin Liu, Tomi Tenhunen, Heidi Kuusniemi, and Yuwei Chen. Inquiry-based bluetooth indoor positioning via RSSI probability distributions. In *2010 Second International Conference on Advances in Satellite and Space Communications*, pages 151–156. IEEE, 2010.

18. Lionel M Ni, Yunhao Liu, Yiu Cho Lau, and Abhishek P Patil. Landmarc: indoor location sensing using active RFID. In *Proceedings of the First IEEE International Conference on Pervasive Computing and Communications, 2003.(PerCom 2003).*, pages 407–415. IEEE, 2003.

19. Jaewoo Chung, Matt Donahoe, Chris Schmandt, Ig-Jae Kim, Pedram Razavai, and Micaela Wiseman. Indoor location sensing using geo-magnetism. In *Proceedings of the 9th international conference on mobile systems, applications, and services*, pages 141–154. ACM, 2011.

20. Janne Haverinen and Anssi Kemppainen. A global self-localization technique utilizing local anomalies of the ambient magnetic field. In *2009 IEEE International Conference on Robotics and Automation*, pages 3142–3147. IEEE, 2009.

21. Gerardine Immaculate Mary and V Prithiviraj. Test measurements of improved UWB localization technique for precision automobile parking. In *2008 International Conference on Recent Advances in Microwave Theory and Applications*, pages 550–553. IEEE, 2008.

22. Mu Zhou, Qiao Zhang, Zengshan Tian, Kunjie Xu, Feng Qiu, and Haibo Wu. IMLOURS: Indoor mapping and localization using time-stamped WLAN received signal strength. In *2015 IEEE Wireless Communications and Networking Conference (WCNC)*, pages 1817–1822. IEEE, 2015.

23. Anastasios Arvanitopoulos, John Gialelis, and Stavros Koubias. Energy efficient indoor localization utilizing bt 4.0 strapdown inertial navigation system. In *Proceedings of the 2014 IEEE Emerging Technology and Factory Automation (ETFA)*, pages 1–5. IEEE, 2014.

24. Angela Song-Ie Noh, Woong Jae Lee, and Jin Young Ye. Comparison of the mechanisms of the Zigbee's indoor localization algorithm. In *2008 Ninth ACIS International Conference on Software Engineering, Artificial Intelligence, Networking, and Parallel/Distributed Computing*, pages 13–18. IEEE, 2008.

25. Mohsen Kavehrad. Sustainable energy-efficient wireless applications using light. *IEEE Communications Magazine*, 48(12):66–73, 2010.

26. JHY Nah, R Parthiban, and MH Jaward. Visible light communications localization using TDOA-based coherent heterodyne detection. In *2013 IEEE 4th international conference on photonics (ICP)*, pages 247–249. IEEE, 2013.

27. Swook Hann, Jung-Hun Kim, Soo-Yong Jung, and Chang-Soo Park. White LED ceiling lights positioning systems for optical wireless indoor applications. In *36th European Conference and Exhibition on Optical Communication*, pages 1–3. IEEE, 2010.

28. Panarat Cherntanomwong and Wisarut Chantharasena. Indoor localization system using visible light communication. In *2015 7th International Conference on Information Technology and Electrical Engineering (ICITEE)*, pages 480–483. IEEE, 2015.

29. Chinnapat Sertthin, Emiko Tsuji, Masao Nakagawa, Shigeru Kuwano, and Kazuji Watanabe. A switching estimated receiver position scheme for visible light based indoor positioning system. In *2009 4th International Symposium on Wireless Pervasive Computing*, pages 1–5. IEEE, 2009.

30. Muhammad Yasir, Siu-Wai Ho, and Badri N Vellambi. Indoor positioning system using visible light and accelerometer. *Journal of Lightwave Technology*, 32(19):3306–3316, 2014.

31. Wenjun Gu, Weizhi Zhang, Jin Wang, MR Amini Kashani, and Mohsen Kavehrad. Three dimensional indoor positioning based on visible light with Gaussian mixture sigma-point particle filter technique. In *Broadband Access Communication Technologies IX*, volume 9387, page 93870O. International Society for Optics and Photonics, 2015.

32. Jay Farrell and Matthew Barth. *The global positioning system and inertial navigation*, volume 61. McGraw-Hill, New York, 1999.

4 Visible Light Positioning Systems

In this chapter, we will investigate 2-D and 3-D positioning systems based on visible light communication. Two different channel access methods along with 2-D and 3-D positioning algorithms will be discussed. Finally, we evaluate the accuracy of the positioning systems in 2-D and 3-D scenarios.

4.1 SYSTEM CONFIGURATION AND CHANNEL ACCESS METHOD

Fig. 4.1 shows a typical optical positioning system model, where the yellow dots represent light emitting diodes (LEDs) installed on a ceiling. Throughout this chapter, we assume a symmetrical cell of room with size of 6m × 6m with height of 3.5 m to analyze the positioning performance. The red pyramid in Fig. 4.1 is a receiver assumed to be located at a height of 1.2 m. In this configuration, the four LED coordinates are (2m, 2m), (2m, 4m), (4m, 2m) and (4m, 4m). Each transmitter is assigned with a unique identification (ID) code to denote their coordinates. A driver circuit is used to modulate the signal in on-off keying (OOK) format. Theoretically, only three LEDs are enough for the coordinate estimation. However, considering that the traditional lighting bulbs are installed in the square layout, the four-LED cell design will not change the basic configuration of an LED illumination system.

4.1.1 TIME DIVISION MULTIPLE ACCESS

Since all LED bulbs transmit their coordinate information independently, the signals will interfere with one another in the air and cannot be retrieved correctly. Therefore, channel access methods should be used in a positioning system. Time division multiple access (TDMA) is a commonly used channel access method where all the transmitters have synchronized frames and occupy different time slots in one frame period to send their signals. The frame structure is shown in Fig. 4.2. In visible light communication (VLC) positioning systems, when one LED transmits its ID information, all the other LEDs emit constant light intensity for the illumination purpose [1].

4.1.2 BASIC FRAMED SLOTTED ALOHA

One disadvantage of TDMA is that the synchronization is required, i.e., all the time slots of the transmitters should start at the same time and the receivers should also know the start time increasing the deployment cost. Basic framed slotted ALOHA (BFSA) is proposed as an alternative asynchronous protocol where the transmitters and the receiver do not need to have the same start point of each time slot. In this

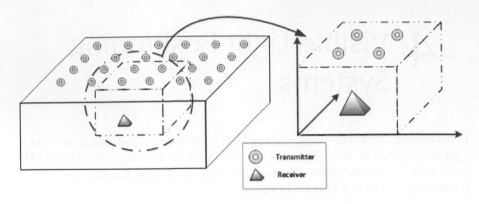

Figure 4.1: System model.

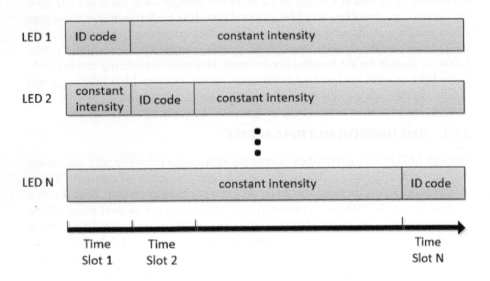

Figure 4.2: Frame structure of the positioning system for one period.

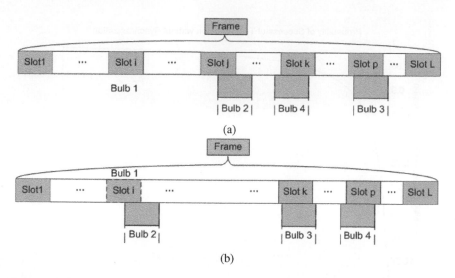

Figure 4.3: Basic framed slotted ALOHA protocol: (a) a successful transmission and (b) a transmission failure since two transmitters select the same slot.

method, a fixed number of time slots are defined in a frame structure for each transmitter. Each transmitter selects one slot randomly in the entire frame to send signals. In BFSA, there are l transmitters to compete for the L slots where $L \geq l$ [2]. Fig. 4.3 demonstrates the working principle of BFSA when $l = 4$. As shown in Fig. 4.3a, if the slots do not overlap with each other, i.e., the receiver can separate the signals without any interference, the transmission is defined as successful. As shown in Fig. 4.3b, if an interference happens, the receiver will fail to distinguish the transmitted signals. Fig. 4.4 demonstrates probability of successful transmission for a different number of slots per frame. Particularly, when L equals 400, the average successful transmission rate is 98.5%. The problems made by the 1.5% failure rate can be compensated with other techniques discussed in the next chapters.

The transmitted data of each LED has a length of 160 bits, with 128 bits of ID data, 8 bits of beginning of flag (BOF), 8 bits of ending of flag (EOF) and another 16 bits of frame correction sequence (FCS). The data structure is shown in Fig. 4.5. Furthermore, cyclic redundancy check (CRC) is applied to ensure the reliability of the transmission.

Despite the advantages of BFSA, the bandwidth is wasted in this method. With 4 LEDs and 400 slots, and if the sampling period is 0.05 s, the required transmission rate is obtained as

$$T_r = \frac{400 \times 160\,\text{bit}}{0.05\,\text{s}} = 1.28\,\text{Mbps}$$

However, only 160 bits/0.05s = 3.2 Kbps is required in TDMA. Although both rates are easily achieved in current VLC technology, TDMA method enables the transmission of other service data.

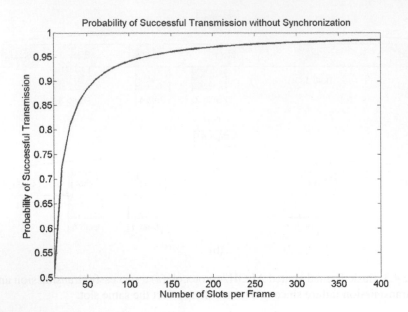

Figure 4.4: Probability of successful transmission versus number of slots per frame.

4.2   POSITIONING ALGORITHM FOR 2-D SCENARIO

As discussed in Chapter 2, the received power from i^{th} LED which is detected by the photodiode (PD) can be obtained in VLC positioning system as

$$P_{r,i} = H(0)P_t = \frac{m+1}{2\pi d_i^2} A\cos^m(\phi) T_s(\psi) g(\psi) \cos(\psi) P_t, \quad i = 1,2,3,4 \qquad (4.1)$$

where P_t is the transmitted power. In the 2-D scenario, the transmitter and the receiver are assumed to be perpendicular to the ceiling resulting in

$$\cos(\phi) = \cos(\psi) = h/d_i \qquad (4.2)$$

where h is the vertical distance between the transmitter and receiver. The distance between the transmitter and the receiver can be estimated as:

$$\hat{d}_i = \sqrt[4]{\frac{(m+1)h^2 A T_s(\psi) g(\psi) P_t}{2\pi P_{r,i}}} \qquad (4.3)$$

Therefore, the estimated horizontal distance $\hat{r}_i$ between the transmitter and receiver is calculated as

$$\hat{r}_i = \sqrt{\hat{d}_i^2 - h^2} = \sqrt{\sqrt{\frac{(m+1)h^2 A T_s(\psi) g(\psi) P_t}{2\pi P_{r,i}}} - h^2} \qquad (4.4)$$

| BOF | LED_IDs | FCS | EOF |

Figure 4.5: Structure of data.

The transmitter's coordinates (x_i, y_i) are decoded from the transmitted ID data, while the receiver's coordinate (x, y) is to be estimated. Based on the lateration technique, the following equation groups are derived:

$$\begin{cases} (x_1 - x)^2 + (y_1 - y)^2 = \hat{r}_1^2 \\ (x_2 - x)^2 + (y_2 - y)^2 = \hat{r}_2^2 \\ (x_3 - x)^2 + (y_3 - y)^2 = \hat{r}_3^2 \\ (x_4 - x)^2 + (y_4 - y)^2 = \hat{r}_4^2 \end{cases} \tag{4.5}$$

Let us subtract the first equation from others to obtain the linear equations as

$$\begin{cases} (x_1 - x_2)x + (y_1 - y_2)y = \left(r_2^2 - r_1^2 - x_2^2 + x_1^2 - y_2^2 + y_1^2 \right)/2 \\ (x_1 - x_3)x + (y_1 - y_3)y = \left(r_3^2 - r_1^2 - x_3^2 + x_1^2 - y_3^2 + y_1^2 \right)/2 \\ (x_1 - x_4)x + (y_1 - y_4)y = \left(r_4^2 - r_1^2 - x_4^2 + x_1^2 - y_4^2 + y_1^2 \right)/2 \end{cases} \tag{4.6}$$

Eq. (4.6) can be expressed in the matrix format as

$$\mathbf{AX} = \mathbf{B} \tag{4.7}$$

where

$$\mathbf{X} = [x, y]^T \tag{4.8}$$

$$\mathbf{A} = \begin{bmatrix} x_2 - x_1 & y_2 - y_1 \\ x_3 - x_1 & y_3 - y_1 \\ x_4 - x_1 & y_4 - y_1 \end{bmatrix} \tag{4.9}$$

$$\mathbf{B} = \frac{1}{2} \begin{bmatrix} \left(r_1^2 - r_2^2 \right) + \left(x_2^2 + y_2^2 \right) - \left(x_1^2 + y_1^2 \right) \\ \left(r_1^2 - r_3^2 \right) + \left(x_3^2 + y_3^2 \right) - \left(x_1^2 + y_1^2 \right) \\ \left(r_1^2 - r_4^2 \right) + \left(x_4^2 + y_4^2 \right) - \left(x_1^2 + y_1^2 \right) \end{bmatrix} \tag{4.10}$$

4.3 LINEAR LEAST SQUARE ESTIMATION

Eq. (4.7) is over-determined, i.e., the number of equations is more than that of variables. In this case, linear least square estimation (LLSE) can be used as a solver, where the estimation minimizes the sum of the square error of each single equation. The squared residuals can be estimated as

$$S_L = \sum_{i=1}^{4} (\hat{x} - x)^2 + (\hat{y} - y)^2 = \left\| \mathbf{B} - \mathbf{A}\hat{\mathbf{X}} \right\|_2^2, \tag{4.11}$$

Table 4.1
Numerical values for system parameters

Parameters	Value
Area (A)	$10^{-4}\,\mathrm{m}^2$
Lambertian mode (m)	1
Field-of-view (FOV) (Ψ_C)	$70°$
$T_s(\psi)$	1
Refraction index of compound parabolic concentrator (CPC) (n)	1.5
Transmitted optical power for logic 1/0 (P_t)	5W/3W

where $\|\mathbf{V}\|_2$ is the Euclidean norm calculated as

$$\|\mathbf{V}\|_2 = \sqrt{v_1^2 + \cdots v_n^2} \tag{4.12}$$

In Eq. (4.11), $\hat{x}$, $\hat{y}$ and $\hat{\mathbf{X}}$ are the estimations of x, y and $\mathbf{X}$, respectively. Eq. (4.11) can be rewritten as

$$S_L = (\mathbf{B} - \mathbf{A}\hat{\mathbf{X}})^T (\mathbf{B} - \mathbf{A}\hat{\mathbf{X}}) = \mathbf{X}^T \mathbf{A}^T \mathbf{A} \mathbf{X} - 2\mathbf{X}^T \mathbf{A}^T \mathbf{B} + \mathbf{B}^T \mathbf{B} \tag{4.13}$$

To minimize S_L, the derivation of Eq. (4.13) is set to zero

$$2\mathbf{A}^T \mathbf{A} \mathbf{X} - 2\mathbf{A}^T \mathbf{B} = 0 \tag{4.14}$$

Finally, $\hat{\mathbf{X}}$ is obtained as

$$\hat{\mathbf{X}} = (\mathbf{A}^T \mathbf{A})^{-1} \mathbf{A}^T \mathbf{B} \tag{4.15}$$

To evaluate the positioning performance in the cell under consideration shown in Fig. 4.1, a pseudo-Poisson path is generated by assuming a user walks across the room defined as a real path. The system parameters used in the simulations are summarized in Table 4.1 where the installation errors of LED bulbs are also considered.

In Fig. 4.6, the real track is shown as the black line, and the red circles represent the estimated values. As it can be seen, only small deviations exist, and the root mean square (RMS) error is just 0.0186 m. To further assess the positioning performance and demonstrate how the system delivers a certain level of service quality, 95% confidence interval error is calculated. Cumulative distribution function (CDF) of the positioning errors is shown in Fig. 4.7, and the 95% confidence interval line is marked out. Therefore, most of the positioning errors are within 0.044 m.

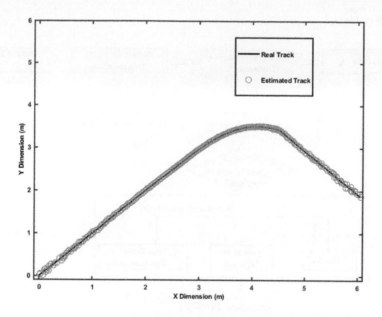

Figure 4.6: Estimated receiver track.

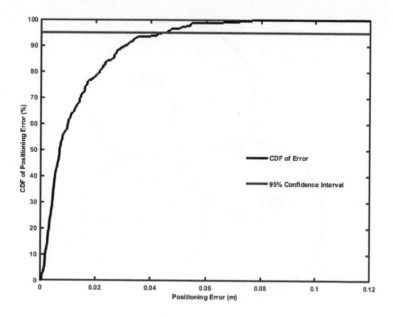

Figure 4.7: CDF of the positioning errors.

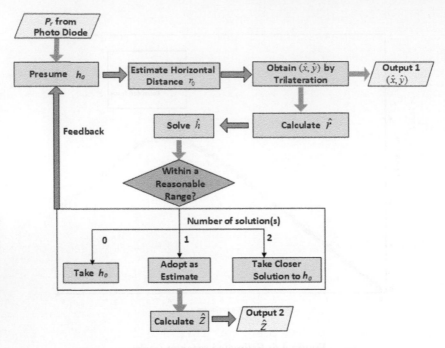

Figure 4.8: Flow diagram of 3-D positioning algorithms.

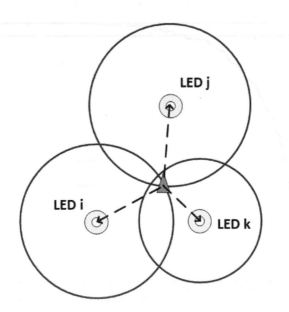

Figure 4.9: Principle of trilateration.

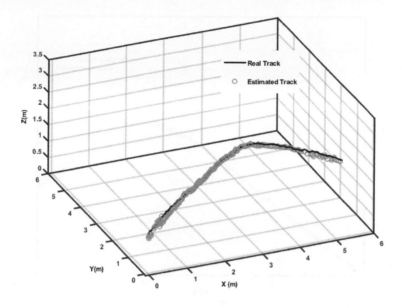

Figure 4.10: 3-D Positioning result.

4.4   POSITIONING ALGORITHM FOR A 3-D SCENARIO

In real scenarios, a user cannot always hold the receiver at the same height level. Even in the robotic applications, the floor cannot always be flat. Therefore, a 3-D positioning algorithm needs to be proposed to address the existence of vertical variations. In order to derive d_i, all the parameters in Eq. (4.3) must be known. However, in the 3-D scenario, the height, h, is not available. To address this issue, a prediction based on the last estimate is made on the height of the receiver as $h^{(0)}$. By substituting $h^{(0)}$ into Eq. (4.3), we obtain

$$\hat{d}_i^{(0)} = \sqrt[4]{\frac{(m+1)\,h^{(0)^2}AT_s\,(\psi)\,g\,(\psi)\,P_t}{2\pi P_{r,i}}} \tag{4.16}$$

The horizontal distance is then estimated as

$$\hat{r}_i^{(0)} = \left(\sqrt{\frac{(m+1)\,h^{(0)^2}AT_s\,(\psi)\,g\,(\psi)\,P_t}{2\pi P_{r,i}}} - h^{(0)2}\right)^{1/2} \tag{4.17}$$

For initializing $h^{(0)}$ for the first estimate, a Gaussian random variable is generated whose mean is the typical hand height of a human being. With the trilateration approach, the horizontal coordinates $[\hat{x}, \hat{y}]$ can be estimated similar to Eq. (4.15), and

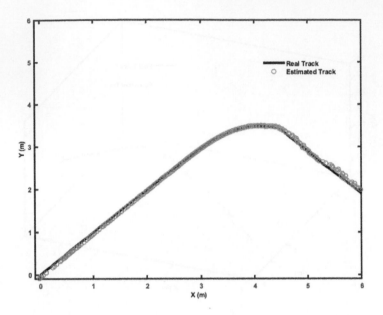

Figure 4.11: Horizontal view of 3-D positioning result.

the horizontal distance can be re-calculated as

$$\hat{r}_i^{(1)} = \sqrt{(x_i - \hat{x})^2 + (y_i - \hat{y})^2} \tag{4.18}$$

When substituting $\hat{r}_i^{(1)}$ into Eq. (4.4), the height is estimated as one solution of

$$\hat{h}^4 + \left(2\hat{r}_i^{(1)2} - C\right)\hat{h}^2 + \hat{r}_i^{(1)4} = 0 \tag{4.19}$$

where C is a constant and obtained as

$$C = \frac{(m+1)AT_s(\psi)g(\psi)P_t}{2\pi P_{r,i}} \tag{4.20}$$

By solving Eq. (4.19), at most two positive $\hat{h}$ are obtained within a reasonable vertical range. If there are two solutions, the one closer to $h^{(0)}$ is selected as $\hat{h}$. If no reasonable solution exists, $h^{(0)}$ will be selected as $\hat{h}$. Finally, the z-coordinate is estimated as

$$\hat{z} = H - \hat{h} \tag{4.21}$$

Fig. 4.8 shows the flow diagram of the 3-D positioning algorithm. This iteration algorithm is applicable because of the principle of trilateration shown in Fig. 4.9. The horizontal position of the receiver is estimated in the intersection of the three

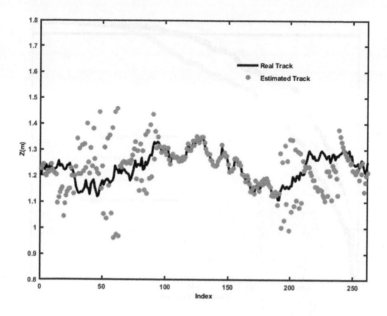

Figure 4.12: Vertical view of 3-D positioning result.

circles. The difference between the presumed height and the actual height only affects the radius of those circles, i.e., all of the circles will expand or shrink at the same time. Therefore, the center of the intersection will remain almost unaffected. However, the change in ratios of these radii is different, which will make the center of the intersection shift a little. This non-uniformity of the noisy received signals induces positioning errors. However, considering that in the 0.05s sampling period, the vertical movement will not be significant, this algorithm still operates properly. The vertical moving distance will be reduced if the sampling period decreases, contributing to higher positioning accuracy.

Eq. (4.10) demonstrates the positioning results in a 3-D scenario where the receiver height varies between 1.0 m to 1.4 m. To demonstrate the result more clearly,

Table 4.2

Errors of the 3-D positioning system (m)

	Horizontal Error	Vertical Error	Total Error
RMS error	0.0735	0.0638	0.0974
95% confidence interval	0.1618	0.1333	0.1977

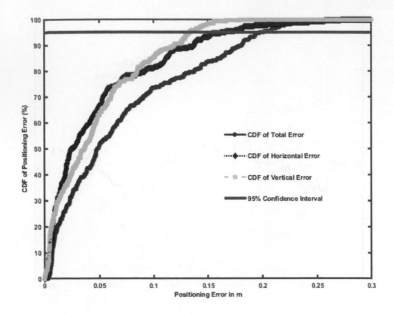

Figure 4.13: CDF curves of the positioning errors.

horizontal and vertical views are plotted separately in Figs. 4.11 and 4.12. Although the vertical estimates have some variations, the horizontal estimates follow the real track very well. Fig. 4.13 shows the CDF of the total, horizontal and vertical positioning errors. The positioning results are summarized in Table 4.2, where the RMS error is just 0.0974 m and 95% of the errors are within 0.1977 m.

SUMMARY

In this chapter, the link layer issues were covered. Channel access methods are addressed. TDMA has the synchronization requirement, but the bandwidth usage is efficient. With BFSA, deployment cost is decreased with no synchronization requirement but bandwidth is wasted. The positioning algorithm is based on the received signal strength (RSS) information detected from PD. Signal attenuation is calculated so that the distance between the transmitters and receiver is estimated. The transmitter's coordinates are obtained by decoding the ID signals. These transmitters act as the center of several different circles and the calculated distances are the radius of these circles. The receiver position is finally estimated as in the intersection of these circles. For the 3-D scenario, positioning is realized by firstly making an assumption on the vertical coordinates and then feeding the initial estimates back to the original equation for a final estimate.

REFERENCES

1. Zhou Zhou, Mohsen Kavehrad, and Peng Deng. Indoor positioning algorithm using light-emitting diode visible light communications. *Optical Engineering*, 51(8):085009, 2012.
2. Weizhi Zhang, MI Sakib Chowdhury, and Mohsen Kavehrad. Asynchronous indoor positioning system based on visible light communications. *Optical Engineering*, 53(4):045105, 2014.

5 Filtering Techniques in Positioning

Filtering techniques are used to further improve the positioning performance. In the positioning system, outliers may appear considering the basic framed slotted ALOHA (BFSA) failure case mentioned in Chapter 4. These large deviations heavily affect the positioning performance for several sequential estimates. Filtering techniques can mitigate these effects.

In this chapter, we review three filtering techniques utilized in visible light communication (VLC) positioning systems, namely Kalman filter, particle filter, and Gaussian mixture sigma-point particle filter (GM-SPPF).

5.1 KALMAN FILTER

In statistics and control theory, Kalman filtering, also known as linear quadratic estimation (LQE), is an algorithm that uses a series of measurements observed over time, containing statistical noise and other inaccuracies, and produces estimates of unknown variables that tend to be more accurate than those based on a single measurement alone, by estimating a joint probability distribution over the variables for each timeframe. The filter is named after Rudolf E. Kalman, one of the primary developers of its theory.

The Kalman filter has numerous applications in technology. A common application is for guidance, navigation, and control of vehicles, particularly aircraft and spacecraft. Furthermore, the Kalman filter is a widely applied concept in time series analysis used in fields such as signal processing and econometrics.

Kalman filters also are one of the main topics in the field of robotic motion planning and control, and they are sometimes included in trajectory optimization. The Kalman filter also works for modeling the central nervous system control of movement. Due to the time delay between issuing motor commands and receiving sensory feedback, use of the Kalman filter supports a realistic model for making estimates of the current state of the motor system and issuing updated commands.

In 1960, Rudolph Kalman published his well-understood recursive solution for the discrete-data linear filtering problem [1]. This filter is named after him as Kalman filter. Since then, Kalman filter has been largely researched and wildly used in many areas, as described.

5.1.1 DERIVATION OF KALMAN FILTER

Mathematically speaking, Kalman filter is formed by a series of recursive equations that provide an efficient way to estimate the state of an evolving random process so that its estimated variance can be minimized. Kalman filter can be used to estimate the previous and the current states, or even the future states of the process, even in the absence of any knowledge of the model.

For the system under consideration, we use a discrete Kalman filter to estimate the state $\mathbf{S} \in \mathbf{R}^n$ by assuming the system follows a linear process [2, 3]

$$\mathbf{S}_k = \mathbf{A}\mathbf{S}_{k-1} + \mathbf{B}\mathbf{u}_k + \mathbf{w}_{k-1} \qquad (5.1)$$

where $\mathbf{A}$ is state transition vector, $\mathbf{B}$ is the input parameter on the input vector $\mathbf{u}_k$, and $\mathbf{w}_{k-1}$ is the process noise. The observation vector, $\mathbf{m}_k \in \mathbf{R}^m$, depends on the current state as

$$\mathbf{m}_k = \mathbf{H}\mathbf{S}_k + \mathbf{v}_k \qquad (5.2)$$

where $\mathbf{H}$ is the observation parameters and $\mathbf{v}_k$ is the observation noise. $\mathbf{w}_k$ and $\mathbf{v}_k$ are assumed to follow Gaussian distribution and independent, i.e., $\mathbf{w}_k \sim N(\mathbf{0}, \mathbf{Q}_k)$ and $\mathbf{v}_k \sim N(\mathbf{0}, \mathbf{R}_k)$ where $\mathbf{Q}_k$ is the process noise covariance and $\mathbf{R}_k$ is the measurement noise covariance.

At step k, $\hat{\mathbf{S}}_k^- \in \mathbf{R}^n$ is the priori state estimation, and $\hat{\mathbf{S}}_k \in \mathbf{R}^n$ is the posteriori state estimation after the measurement $\mathbf{m}_k$ is given. Therefore, the priori and posteriori estimate errors are

$$\mathbf{e}_k^- = \mathbf{S}_k - \hat{\mathbf{S}}_k^- \qquad (5.3a)$$

$$\mathbf{e}_k = \mathbf{S}_k - \hat{\mathbf{S}}_k \qquad (5.3b)$$

The posteriori estimate is based on a linear blending of the noisy measurement and the priori estimate

$$\hat{\mathbf{S}}_k = \hat{\mathbf{S}}_k^- + \mathbf{K}_k \left(\mathbf{m}_k - \mathbf{H}\hat{\mathbf{S}}_k^- \right) \qquad (5.4)$$

In Eq. (5.4), $\mathbf{K}_k$ is considered as the blending factor. In order to find the appropriate $\mathbf{K}_k$ that generates an optimal updated estimate, minimum mean-square error is used as the performance criterion. For the posteriori estimate, the expression for the error covariance is given by

$$\mathbf{P}_k = E\left[\mathbf{e}_k\mathbf{e}_k^T\right] = E\left[\left(\mathbf{S}_k - \hat{\mathbf{S}}_k\right)\left(\mathbf{S}_k - \hat{\mathbf{S}}_k\right)^T\right] \qquad (5.5)$$

$$= E\left[\left(\mathbf{S}_k - \hat{\mathbf{S}}_k^- - \mathbf{K}_k\left(\mathbf{H}\mathbf{S}_k + \mathbf{v}_k - \mathbf{H}\hat{\mathbf{S}}_k^-\right)\right)\left(\mathbf{S}_k - \hat{\mathbf{S}}_k^- - \mathbf{K}_k\left(\mathbf{H}\mathbf{S}_k + \mathbf{v}_k - \mathbf{H}\hat{\mathbf{S}}_k^-\right)\right)^T\right]$$

The priori error, i.e., $\left(\mathbf{S}_k - \hat{\mathbf{S}}_k^-\right)$, is uncorrelated with the measurement error $\mathbf{v}_k$. Therefore, Eq. (5.5) can be expressed as:

$$\mathbf{P}_k = E\left[\left(\mathbf{S}_k - \hat{\mathbf{S}}_k\right)(\mathbf{I} - \mathbf{K}_k\mathbf{H})\left(\left(\mathbf{S}_k - \hat{\mathbf{S}}_k\right)(\mathbf{I} - \mathbf{K}_k\mathbf{H})\right)^T\right] + E\left[(\mathbf{K}_k\mathbf{v}_k)(\mathbf{K}_k\mathbf{v}_k)^T\right]$$

$$= (\mathbf{I} - \mathbf{K}_k\mathbf{H})\mathbf{P}_k^-(\mathbf{I} - \mathbf{K}_k\mathbf{H})^T + \mathbf{K}_k\mathbf{R}_k\mathbf{K}_k^T \qquad (5.6)$$

$$= \mathbf{P}_k^- - \mathbf{K}_k\mathbf{H}\mathbf{P}_k^- - \mathbf{P}_k^-\mathbf{H}^T\mathbf{K}_k^T + \mathbf{K}_k\left(\mathbf{H}\mathbf{P}_k^-\mathbf{H}^T + \mathbf{P}_k\right)\mathbf{K}_k^T$$

where $\mathbf{P}_k^-$ is the estimated priori error covariance and given by

$$\mathbf{P}_k^- = E\left[\mathbf{e}_k^- \mathbf{e}_k^{-T}\right] = E\left[\left(\mathbf{S}_k - \hat{\mathbf{S}}_k^-\right)\left(\mathbf{S}_k - \hat{\mathbf{S}}_k^-\right)^T\right] \qquad (5.7)$$

The trace of $\mathbf{P}_k$ is the sum of the mean square errors. Considering that the individual error is minimized when the mean-square error is minimized, we differentiate the trace of $\mathbf{P}_k$ and then set it to zero

$$\frac{d\left(\mathrm{tr}\left(\mathbf{P}_k\right)\right)}{d\mathbf{K}_k} = -2\left(\mathbf{H}_k\mathbf{P}_k^-\right)^T + 2\mathbf{K}_k\left(\mathbf{H}_k\mathbf{P}_k^-\mathbf{H}_k^T + \mathbf{R}_k\right) = 0 \qquad (5.8)$$

Therefore,

$$\mathbf{K}_k = \mathbf{P}_k^-\mathbf{H}^T\left(\mathbf{H}\mathbf{P}_k^-\mathbf{H}^T + \mathbf{R}\right)^{-1} \qquad (5.9)$$

Inserting Eq. (5.9) in Eq. (5.6), we obtain

$$\mathbf{P}_k = \mathbf{P}_k^- - \mathbf{P}_k^-\mathbf{H}^T\left(\mathbf{H}\mathbf{P}_k^-\mathbf{H}^T + \mathbf{R}_k\right)^{-1}\mathbf{H}\mathbf{P}_k^- = (\mathbf{I} - \mathbf{K}_k\mathbf{H})\mathbf{P}_k^- \qquad (5.10)$$

In addition, we can obtain

$$\mathbf{e}_{k+1}^- = \mathbf{S}_{k+1} - \hat{\mathbf{S}}_{k+1}^- = \mathbf{A}\mathbf{S}_k + \mathbf{w}_k - \mathbf{A}\hat{\mathbf{S}}_k = \mathbf{A}\mathbf{e}_k + \mathbf{w}_k \qquad (5.11)$$

$$\mathbf{P}_{k+1}^- = E\left[\left(\mathbf{A}\mathbf{S}_k + \mathbf{w}_k - \mathbf{A}\hat{\mathbf{S}}_k\right)\left(\mathbf{A}\mathbf{S}_k + \mathbf{w}_k - \mathbf{A}\hat{\mathbf{S}}_k\right)^T\right] = \mathbf{A}\mathbf{P}_k\mathbf{A} + \mathbf{Q}_k \qquad (5.12)$$

In summary, Kalman filtering is done in two phases as shown in Fig. 5.1. In the time update phase, the current state $\hat{\mathbf{S}}_k^-$ and the covariance $\mathbf{P}_k^-$ are predicted based on the previous information. When $\mathbf{m}_k$ is available, the measurement update phase begins. Kalman gain $\mathbf{K}_k$ is computed and the posterior state $\hat{\mathbf{S}}_k$ and the covariance $\mathbf{P}_k$ are updated.

5.1.2 KALMAN FILTER IN 2-D AND 3-D SCENARIOS

In a proposed 2-D scenario, the state vector is expressed as $\mathbf{S}_k = \left[x_k, y_k, v_{xk}, v_{yk}\right]$ in which $[x_k, y_k]$ represents the coordinates, $\mathbf{u}_k$ is 0. Parameters $\mathbf{v}_{xk}$ and $\mathbf{v}_{yk}$ are the movement speeds in the x and y directions, respectively. $\mathbf{A}$, $\mathbf{H}$, $\mathbf{Q}$ and $\mathbf{R}$ are expressed as:

$$\mathbf{A} = \begin{bmatrix} 1 & 0 & \Delta t & 0 \\ 0 & 1 & 0 & \Delta t \\ 0 & 0 & 1 & 0 \\ 0 & 0 & 0 & 1 \end{bmatrix} \qquad (5.13)$$

$$\mathbf{H} = \begin{bmatrix} 1 & 0 & \Delta t & 0 \\ 0 & 1 & 0 & \Delta t \end{bmatrix} \qquad (5.14)$$

$$\mathbf{Q} = \begin{bmatrix} \sigma_{Px}^2 & 0 & 0 & 0 \\ 0 & \sigma_{Py}^2 & 0 & 0 \\ 0 & 0 & \sigma_{Vx}^2 & 0 \\ 0 & 0 & 0 & \sigma_{Yx}^2 \end{bmatrix} \qquad (5.15)$$

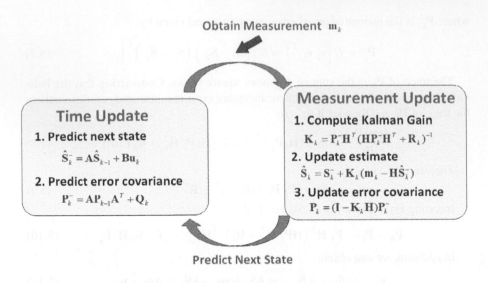

Figure 5.1: Principle of Kalman filter.

$$\mathbf{R} = \begin{bmatrix} \sigma_x^2 & 0 \\ 0 & \sigma_y^2 \end{bmatrix} \qquad (5.16)$$

where Δt is the sampling period, σ_{Px}^2 and σ_{Py}^2 are the variances of the receiver movement initialized as 0.1. σ_{Vx}^2 and σ_{Vy}^2 are the variances of receiver speed initialized as 0.5, and σ_x^2 and σ_y^2 are the measurement variances decided by the photodiode (PD). $\mathbf{K}_k$ is a 4×4 matrix which is initialized as

$$\mathbf{K}_0 = \begin{bmatrix} 2 & 0 & 0 & 0 \\ 0 & 2 & 0 & 0 \\ 0 & 0 & 4 & 0 \\ 0 & 0 & 0 & 4 \end{bmatrix} \qquad (5.17)$$

In the 3-D scenario, the receiver vertical movement is independent of horizontal movement. Therefore, the filtering process is divided into horizontal and vertical phases. Kalman filter is applied on the horizontal estimation first, and then the filtered results are fed back into the 3-D positioning algorithm to obtain the z-coordinate. Fig. 5.2 shows a flow diagram of Kalman filter for 3-D scenario that is summarized as follows:

1. Initialize the $\hat{\mathbf{S}}_0 = [\hat{x}_0, \hat{y}_0]$ based on the first horizontal measurement from the light emitting diode (LED)-positioning algorithm as mentioned in Chapter 4.
2. Make the prediction on the current state vector $\hat{\mathbf{S}}_k^- = [\hat{\mathbf{x}}_k^-, \hat{\mathbf{y}}_k^-]$ based on the previous position $\hat{\mathbf{S}}_{k-1} = [\hat{\mathbf{x}}_{k-1}, \hat{\mathbf{y}}_{k-1}]$ and the movement velocity. $\mathbf{P}_k^-$ is also predicted based on the last estimate.

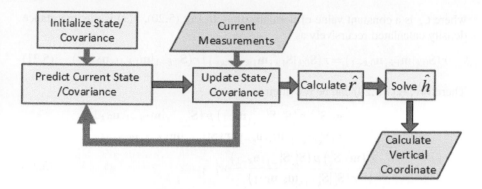

Figure 5.2: Flow diagram of Kalman Filter for 3-D scenario.

3. Obtain the measurement value $\mathbf{m}_k = [\hat{x}, \hat{y}]$ as mentioned in Chapter 4.
4. Update $\hat{\mathbf{S}}_k = [\hat{x}_k, \hat{y}_k]$ and $\mathbf{P}_k$ with the measurement value.
5. Insert $[\hat{x}_k, \hat{y}_k]$ into the 3-D positioning algorithm to obtain $\hat{r}_k$, $\hat{h}_k$ and $\hat{z}_k$.

5.2 PARTICLE FILTER

One drawback of Kalman filtering is that it is based on the assumption that the state follows Gaussian distribution and the modeling process is linear. For a nonlinear and non-Gaussian situation, a particle filter is proposed, based on Monte Carlo estimation [4].

5.2.1 PRINCIPLE OF PARTICLE FILTER

A set of weighted samples at k^{th} step has the posterior distribution as

$$(\mathbf{S}_k|\mathbf{m}_{1:k}, \mathbf{u}_{0:k-1}) \approx \sum_{i=1}^{N} w_k^i \delta\left(\mathbf{S}_k - \mathbf{S}_k^i\right) \qquad (5.18)$$

where N is the number of samples, and i is the sample index. $\delta(.)$ is the Dirac delta function and w_k^i is the weight of the ith sample. Based on the probability theory, the posterior distribution can be estimated as

$$p\left(\mathbf{S}_{0:k}|\mathbf{m}_{1:k}, \mathbf{u}_{0:k-1}\right) = \frac{p\left(\mathbf{m}_k|\mathbf{S}_k\right) p\left(\mathbf{S}_k|\mathbf{S}_{k-1}, \mathbf{u}_{k-1}\right) p\left(\mathbf{S}_{0:k-1}|\mathbf{m}_{1:k-1}, \mathbf{u}_{0:k-2}\right)}{p\left(\mathbf{m}_k|\mathbf{m}_{1:k-1}, \mathbf{u}_{0:k-1}\right)} \qquad (5.19)$$

The normalized weight of each particle is

$$w_k^i = C_k \frac{p\left(\mathbf{S}_{0:k}^i|\mathbf{m}_{1:k}, \mathbf{u}_{0:k-1}\right)}{r\left(\mathbf{S}_{0:k}^i|\mathbf{m}_{1:k}, \mathbf{u}_{0:k-1}\right)} \qquad (5.20)$$

$$= C_k \frac{p\left(\mathbf{m}_k|\mathbf{S}_k^i\right) p\left(\mathbf{S}_k^i|\mathbf{S}_{k-1}^i, \mathbf{u}_{k-1}\right) p\left(\mathbf{S}_{0:k-1}^i|\mathbf{m}_{1:k-1}, \mathbf{u}_{0:k-2}\right)}{r\left(\mathbf{S}_{0:k}^i|\mathbf{m}_{1:k}, \mathbf{u}_{0:k-1}\right)}$$

where C_k is a constant value of normalization. In Eq. (5.20), $r(x)$ is the importance density calculated recursively as

$$r(\mathbf{S}_{0:k}|\mathbf{m}_{1:k},\mathbf{u}_{0:k-1}) = r(\mathbf{S}_{0:k}|\mathbf{S}_{1:k},\mathbf{m}_{1:k},\mathbf{u}_{0:k-1})\,r(\mathbf{S}_{0:k-1}|\mathbf{m}_{1:k-1},\mathbf{u}_{0:k-2}) \qquad (5.21)$$

Therefore, Eq. (5.20) can be rewritten as

$$
\begin{aligned}
w_k^i &= C_k \frac{p\left(\mathbf{m}_k|\mathbf{S}_k^i\right) p\left(\mathbf{S}_k^i|\mathbf{S}_{k-1}^i,\mathbf{u}_{k-1}\right) p\left(\mathbf{S}_{0:k-1}^i|\mathbf{m}_{1:k-1},\mathbf{u}_{0:k-2}\right)}{r\left(\mathbf{S}_k^i|\mathbf{S}_{k-1}^i,\mathbf{m}_k,\mathbf{u}_{k-1}\right) r\left(\mathbf{S}_{0:k-1}^i|\mathbf{m}_{1:k-1},\mathbf{u}_{0:k-2}\right)} \\
&= C_k \frac{p\left(\mathbf{m}_k|\mathbf{S}_k^i\right) p\left(\mathbf{S}_k^i|\mathbf{S}_{k-1}^i,\mathbf{u}_{k-1}\right)}{r\left(\mathbf{S}_k^i|\mathbf{S}_{k-1}^i,\mathbf{m}_k,\mathbf{u}_{k-1}\right)} w_{k-1}^i
\end{aligned}
\qquad (5.22)
$$

If $[r(\mathbf{S}_k|\mathbf{S}_{k-1},\mathbf{m}_k,\mathbf{u}_{k-1}) = p(\mathbf{S}_k|\mathbf{S}_{k-1},\mathbf{u}_{k-1})$, then Eq. (5.22) is simplified as

$$w_k^i = C_k p\left(\mathbf{m}_k|\mathbf{S}_k^i\right) w_{k-1}^i \qquad (5.23)$$

The state of kth step is ultimately estimated as

$$\hat{\mathbf{S}}_k = \sum_{i=1}^{N} w_k^i \mathbf{S}_k^i \qquad (5.24)$$

The main disadvantage of particle filtering is that the weight of some particles can keep on increasing during the iteration process and dominate that for all the other particles. A resampling process making the operation of a particle filter more stable can overcome this problem. The weights of N particles are reset to the initial value as $1/N$, and the posterior probability is

$$p(\mathbf{S}_k|\mathbf{m}_{1:k},\mathbf{u}_{0:k-1}) \approx \frac{1}{N}\sum_{i=1}^{N}\delta\left(\mathbf{S}_k-\mathbf{S}_k^i\right) \qquad (5.25)$$

5.2.2 PARTICLE FILTER IN 2-D AND 3-D SCENARIOS

For a 2-D system, a particle filter is applied as follows:

1. Use the first measurement to initialize N particles (x_0^i,y_0^i) with identical weight of $w_0^i = 1/N$ where N equals 500.
2. Calculate (x_k^i,y_k^i) from (x_{k-1}^i,y_{k-1}^i) which is based on the movement distance and direction.
3. Obtain the measurement value as $\mathbf{m}_k = [m_{kx},m_{ky}]$, then update the weight as

$$p\left(\mathbf{m}_k|\mathbf{S}_k^i\right) = \exp\left(-\left(\left(m_{kx}-x_k^i\right)^2 + \left(m_{ky}-y_k^i\right)^2\right)/\sigma^2\right) \qquad (5.26)$$

$$w_k^i = w_{k-1}^i p\left(\mathbf{m}_k|\mathbf{S}_k^i\right) \qquad (5.27)$$

4. Calculate $J = 1/\sum_{i=1}^{N}\left(w_k^i\right)^2$. If it is larger than a predetermined threshold value, the resampling process is conducted by setting $w_k^i = 1/N$.

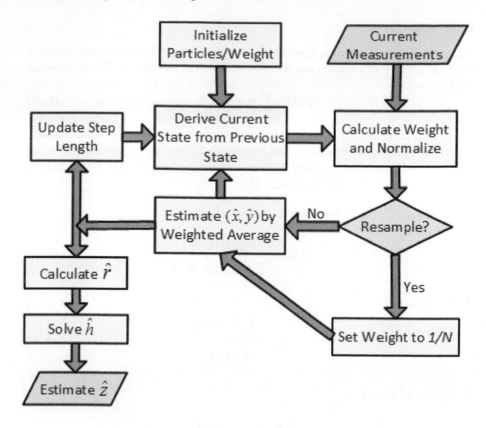

Figure 5.3: Flow diagram of particle filtering.

5. Estimate the target coordinates as $(\hat{x}_k, \hat{y}_k) = \sum\limits_{i=1}^{N} \left(w_k^i x_k^i, w_k^i y_k^i \right)$.

6. Update the step length for each particle and generate new random movement directions.

For a 3-D scenario, after the horizontal coordinates $(\hat{x}_k, \hat{y}_k)$ are processed with particle filter, we insert it into the 3-D algorithm as mentioned in Chapter 4 to obtain $\hat{r}_k$. $\hat{h}_k$ and $\hat{z}_k$ are then calculated. The flow diagram of particle filter is shown in Fig. 5.3.

5.3 GAUSSIAN MIXTURE SIGMA-POINT PARTICLE FILTER

GM-SPPF refers to a set of particle filters which make use of Gaussian mixture model (GMM). GMM methods approximate a probability density function (PDF) represented by a weighted sum of Gaussian components.

5.3.1   PRINCIPLE

The posterior distribution is approximated with a set of Gaussian components to limit the number of particles and decrease the computational cost. The distribution is expressed as

$$p(x) = \sum_i \alpha_i g(x; \mu_i, \Sigma_i) \tag{5.28}$$

where α_i is the normalized weight of each component, and $g(x; \mu_i, \Sigma_i)$ is the Gaussian density component with mean μ_i, and covariance Σ_i. For the sake of simplicity, we can assume Σ_i equals $\sigma_i^2 I$, where I denotes an identity matrix. The GMM PDF is constructed through the following steps [5]:

1. α_i, μ_i and σ_i^2 is initialized with K-means method.
2. Calculate

$$\beta_i(x) = \frac{\alpha_i g\left(x; \mu_i, \sigma_i^2\right)}{\sum_j \alpha_j g\left(x; \mu_j, \sigma_j^2\right)} \tag{5.29}$$

3. Evaluate

$$\mu_i = \frac{\sum_{j=1}^{n} \beta_i(x_j) x_j}{\sum_{j=1}^{n} \beta_i(x_j)} \tag{5.30}$$

4. Update σ_i^2 and α_i as

$$\sigma_i^2 = \frac{\sum_{j=1}^{n} \beta_i(x_j)(x_j - \mu_i)^T (x_j - \mu_i)}{\sum_{j=1}^{n} \beta_i(x_j)} \tag{5.31}$$

$$\alpha_i = \frac{1}{n} \sum_{j=1}^{n} \beta_i(x_j) \tag{5.32}$$

5. Estimate $J = \sum_{i=1}^{n} \ln[\alpha_i g(x; \mu_i, \Sigma_i)]$. Then, compare it to the previous J. If their difference is smaller than a pre-determined threshold, it is considered as a convergence. If not, then go back to Step 2 for another iteration.

After GMM is constructed as the PDF, another technique named as sigma-point approach (SPA) is employed [6]. As shown in Fig. 5.4, a set of sampling points represented by green dots are selected. The coordinates of these points are summarized in Table 5.1.

These selected points propagate through the nonlinear process, and the posterior statics are then evaluated. The weights are updated with the same measurement value as the particle filter and, if necessary, resampling process is applied.

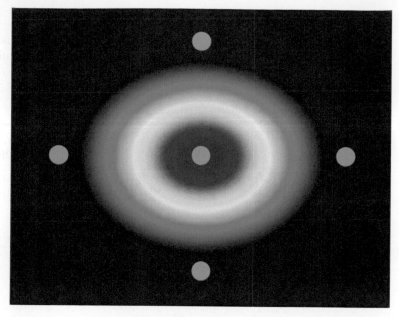

Figure 5.4: An example of the sigma-point approach in a 2-D system.

Table 5.1

Selection of sigma points for each GMM

Position	Sigma-points value
Center	(μ_x, μ_y)
Left	$(\mu_x - \sigma_x, \mu_y)$
Right	$(\mu_x + \sigma_x, \mu_y)$
Up	$(\mu_x, \mu_y + \sigma_y)$
Down	$(\mu_x, \mu_y - \sigma_y)$

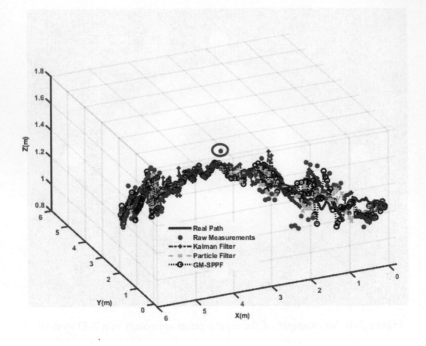

Figure 5.5: 3-D positioning results.

5.3.2 GM-SPPF IN 2-D AND 3-D SCENARIOS

In our system, 50 sigma points are selected for each Gaussian component, and three Gaussian components are applied in total [7]. They propagate through the system function $\mathbf{S}_{k+1} = f(\mathbf{S}_k)$, which relates the next step, $k+1$, to the current step k with moving distance and direction information. The detailed process of applying GM-SPPF is as follows:

1. Initialize N particles $\left(x_0^i, y_0^i\right)$ based on the first measurement value.
2. Cluster N particles into three groups using K-means algorithm.
3. Set up GMM and then sample the sigma points considering five components.
4. When the measurement value is obtained from the PD, the weight is updated using Eq. (5.27) and resampling is applied, if needed.
5. Estimate the target coordinates as $(\hat{x}_k, \hat{y}_k) = \sum\limits_{i=1}^{N} \left(w_k^i x_k^i, w_k^i y_k^i\right)$
6. Update the step length for each particle and generate new random movement directions.

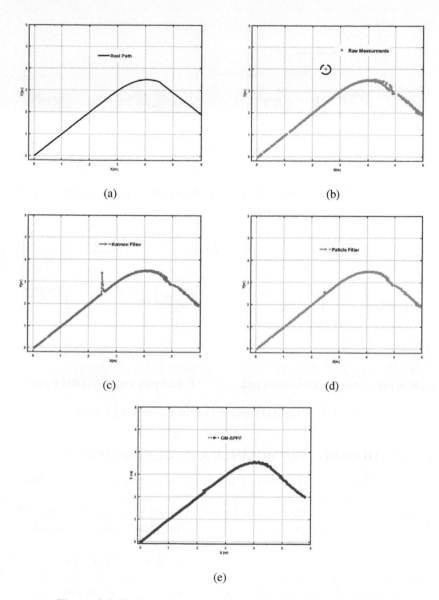

(a)   (b)

(c)   (d)

(e)

Figure 5.6: Horizontal components of the positioning results.

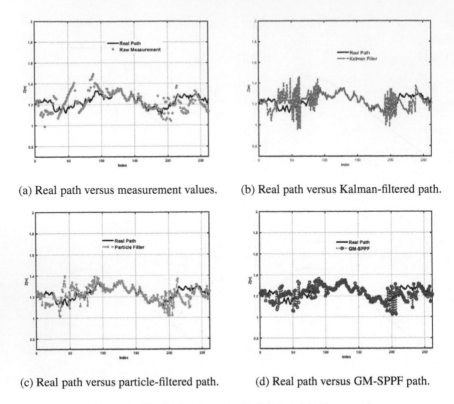

(a) Real path versus measurement values.   (b) Real path versus Kalman-filtered path.

(c) Real path versus particle-filtered path.   (d) Real path versus GM-SPPF path.

Figure 5.7: Vertical component of the positioning results.

5.4   POSITIONING PERFORMANCE AND DISCUSSIONS

Fig. 5.5 shows the positioning results of the 3-D scenario where a large deviation can be seen as cycled out. Real path is shown as the black line and 262 raw measurement values are represented by the red dots. Large deviations may come from the failure of BFSA or the blockage of line-of-sight (LoS) link. The results of Kalman filter, particle filter and GM-SPPF are also shown in this figure.

To demonstrate the results more clearly, horizontal and vertical views were also plotted separately, as shown in Fig. 5.6 and Fig. 5.7. In Fig. 5.6b, the circled red spot shows where the large deviation is located. When the Kalman filter is applied as seen in Fig. 5.6b, estimations still diverge after the large deviation, and it takes several steps before the path converges back to normal again. As shown in Figs. 5.6d and 5.6e, when particle filter and GM-SPPF are applied, the paths after the large deviation are almost unaffected.

As seen in Fig. 5.7, the performance of vertical components is slightly improved by applying filters since they are not directly applied on the vertical path. Furthermore, the measurements are more precise in the center region than at both ends of the path since light intensities are more uniform at the center of the room.

Table 5.2

RMS error of 3-D system with filtering techniques

	Horizontal(m)	Vertical(m)	Total(m)	Time(s)
Raw measurement	0.116	0.086	0.143	NA
Kalman filter	0.084	0.079	0.115	0.46
Particle filter	0.061	0.074	0.096	5.9
GM-SPPF	0.053	0.068	0.086	4.9

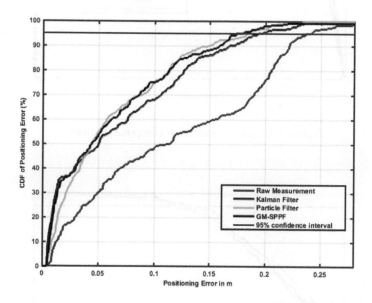

Figure 5.8: Cumulative distribution function (CDF) of 3-D positioning errors.

Table 5.3

95% confidence interval errors of the 3-D system and filtering techniques

	Horizontal(m)	Vertical(m)	Total(m)
Raw measurement	0.1413	0.1733	0.2321
Kalman filter	0.1404	0.1660	0.1931
Particle filter	0.1403	0.1483	0.1876
GM-SPPF	0.1245	0.1177	0.1742

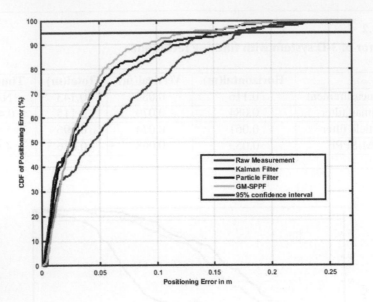

Figure 5.9: CDF of horizontal component errors.

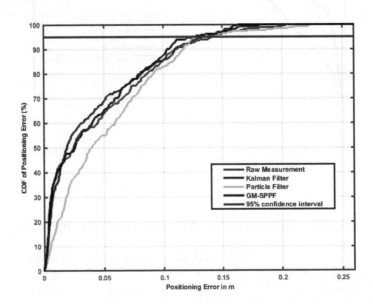

Figure 5.10: CDF of vertical component errors.

The root mean square (RMS) errors are quantitatively summarized in Table 5.2. As it can be clearly seen, an overall accuracy of 0.143 m is achieved, and the particle filter outperforms Kalman filter. In addition, although GM-SPPF slightly improves the performance, the time consumption decreases to 4.9s.

The CDF of overall, horizontal and vertical positioning errors are shown in Figs. 5.8 to 5.10, respectively. The performance improvement is significant after filtering process for both overall and horizontal components, while it is not much noticeable on the vertical components.

The 95% confidence interval errors are also summarized in Table 5.3. We can see most of the errors are within 0.2321 m before the filtering is applied, while they are within 0.1742 m by employing filtering techniques.

SUMMARY

In this chapter three filtering techniques are introduced and applied to remove the large deviations and achieve a high accuracy. Kalman filtering includes two phases that only can be used for linear and Gaussian model. Particle filter, as a Monte Carlo approach, uses a set of particles to approximate the posterior distribution. GM-SPPF uses a set of Gaussian models to represent the particles.

REFERENCES

1. Rudolph Emil Kalman. A new approach to linear filtering and prediction problems. *Journal of basic Engineering*, 82(1):35–45, 1960.
2. Robert Grover Brown, Patrick YC Hwang, et al. *Introduction to random signals and applied Kalman filtering*, volume 3. Wiley New York, 1992.
3. Greg Welch, Gary Bishop, et al. An introduction to the kalman filter. 1995.
4. Seong-hoon Peter Won, Wael William Melek, and Farid Golnaraghi. A kalman/particle filter-based position and orientation estimation method using a position sensor/inertial measurement unit hybrid system. *IEEE Transactions on Industrial Electronics*, 57(5):1787–1798, 2010.
5. Simo Ali-Löytty and Niilo Sirola. Gaussian mixture filters in hybrid positioning. In *Proc. ION GNSS*, pages 562–570, 2009.
6. Rudolph Van Der Merwe, Eric A Wan, et al. Sigma-point kalman filters for integrated navigation. In *Proceedings of the 60th Annual Meeting of the Institute of Navigation (ION)*, pages 641–654, 2004.
7. Wenjun Gu, Weizhi Zhang, Mohsen Kavehrad, and Lihui Feng. Three-dimensional light positioning algorithm with filtering techniques for indoor environments. *Optical Engineering*, 53(10):107107, 2014.

6 Three-Dimensional Positioning Based on Nonlinear Estimation and Multiple Receivers

In Chapter 5, the horizontal positioning performance is improved by employing filtering techniques. However, the vertical positioning performance is not noticeably improved. In this chapter, we first introduce a nonlinear estimation to improve the positioning performance particularly in the vertical direction.

In addition, in the latter parts of this chapter, multi eye receivers with an overall wide field-of-view (FOV) are introduced. These receivers, like a fly-eye, have multiple receive photodiodes (PDs) that provide angle of arrival (AOA) diversity that can be exploited by proper combining of the received signals from different directions of arrival. This configuration, by the narrow FOV of each multi eye, is resilient to multiple received reflected rays that cause dispersion in high rate data streams. This in turn improves the positioning accuracy compared to a single receive-eye configuration.

6.1 3-D POSITIONING BASED ON NONLINEAR ESTIMATION

Thus far, the height of receiver has been assumed to be known, so that the coordinates on the horizontal plane can be calculated. The positioning method in this section includes two stages. First, the height is presumed in the prediction stage. Second, a nonlinear estimation is applied in the correction stage to realize three-dimensional coordinate estimation.

6.1.1 TRUST REGION REFLECTIVE ALGORITHM

Trust region algorithm is a popular solver in the optimization problems [1] which is stated as

$$\min_{x \in \Re^n} f(x)$$
$$\text{s.t.} \quad \begin{matrix} c_i(x) = 0 & i = 1, 2, \cdots m \\ c_i(x) \geq 0 & i = m+1, m+2, \cdots q \end{matrix} \quad (6.1)$$

where $f(x)$ is a nonlinear function. If $m = q = 0$, Eq. (6.1) becomes an unconstrained problem. Assuming that an approximate solution x_k is available at k^{th} iteration, a new point x_{k+1} is found in a trusted region near the current solution. The trust region is enlarged when the current solution fits the problem well; otherwise, it is shrunk. The iteration stops when the convergence condition is satisfied. The most important part of a trust region algorithm is to find the trial step, which can be solved by Levenberg-Marquardt method [2]. This step is expressed by

$$d_k = -\left(\mathbf{J}(x_k)\mathbf{J}(x_k)^T + \lambda_k\mathbf{I}\right)^{-1}\mathbf{J}(x_k)f(x_k) \qquad (6.2)$$

where $\mathbf{J}(x)$ is the Jacobi matrix of $f(x)$ and $\lambda_k \geq 0$ is damping factor adjusted at each iteration.

The general steps are as follows:

1. A trust region is initialized.
2. An approximate model is set up and the trial step, s_k, is found within the trust region.
3. A cost function is applied to update the next trust region and select new points.

For 2-D situation, the problem is stated as:

1. Set up the 2-D trust region sub-problem as

$$\min\left\{\tfrac{1}{2}\mathbf{S}^T\mathbf{H}\mathbf{S} + \mathbf{S}^T\mathbf{g}\right\}$$
$$\text{s.t. } \|\mathbf{DS}\| < \Delta \qquad (6.3)$$

where $\mathbf{H}$ is the Hessian matrix, $\mathbf{g}$ is the gradient of $f(x_{\text{current}})$, $\mathbf{S}$ is the step, $\mathbf{D}$ is a diagonal scaling matrix, and Δ is a positive scalar.

2. Solve Eq. (6.2) to determine $\mathbf{S}$.
3. Set $\mathbf{x} = \mathbf{x} + \mathbf{S}$ when $f(\mathbf{x}+\mathbf{S}) < f(\mathbf{x})$.
4. AdjustΔ.
5. Decide whether the convergence condition is satisfied, if not, go back to step one for the next iteration.

6.1.2   3-D POSITIONING ALGORITHM

When the height information is not available, we set height $h^{(0)}$ based on the previous estimate. In Chapter 4, we learned how to estimate the distance d_i and the horizontal coordinates $\hat{\mathbf{X}}=[\hat{x},\hat{y}]^T$. $\hat{\mathbf{X}}$ is used for initialization which will be discussed later.

For a 3-D positioning scenario, trust region reflective algorithm is also used to find the solution of

$$\min\left\{\tilde{S} = \sum_{i=1}^{4}\left(\sqrt{(x-x_i)^2 + (y-y_i)^2 + (z-z_i)^2} - d_i\right)^2\right\} \qquad (6.4)$$

In this case, an initial value $\hat{\mathbf{X}}^{(0)} = \left[\tilde{x}^{(0)}, \tilde{y}^{(0)}, \tilde{z}^{(0)}\right]$ is first provided and the corresponding $\tilde{S}^{(0)}$ is calculated, where $\hat{\mathbf{X}}$ and $\tilde{z}^{(0)} = z_i - h^{(0)}$. Second, several points surrounding $\hat{\mathbf{X}}^{(0)}$ are selected and their corresponding $\tilde{S}^{(1)}$ are calculated. $\hat{\mathbf{X}}^{(1)}$ is updated with the point that minimizes $\tilde{S}^{(1)}$. After several iterative steps, $\hat{\mathbf{X}}$ is finally obtained when $\tilde{S}$ converges. The obtained $\tilde{z}$ is used to set $h^{(0)}$ at the next sampling. The flow diagram of this algorithm is shown in Fig. 6.1.

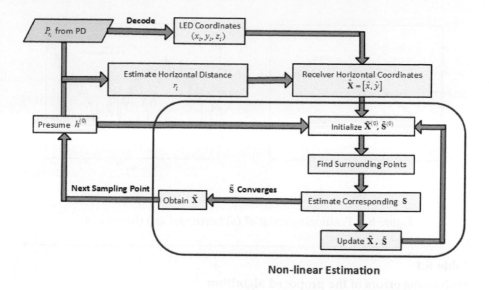

Figure 6.1: Flow diagram of 3-D positioning algorithm.

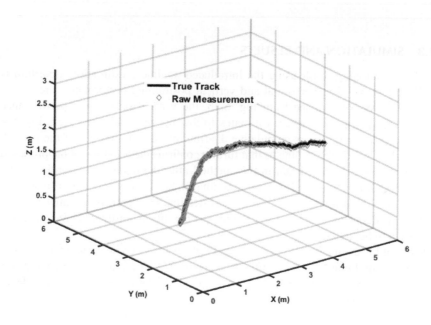

Figure 6.2: Total positioning result.

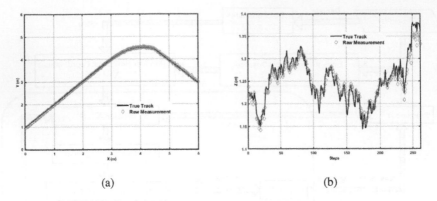

(a)   (b)

Figure 6.3: Positioning result of (a) horizontal and (b) vertical.

Table 6.1

Positioning errors of the proposed algorithm

	Horizontal(m)	Vertical(m)	Total(m)
RMS error	0.0222	0.0120	0.0252
95% Confidence Interval	0.0421	0.0199	0.0432

6.1.3 SIMULATION AND RESULTS

The 3-D positioning result using the introduced nonlinear estimation algorithm is shown in Fig. 6.2. The horizontal and vertical positioning performance are shown in Fig. 6.3. As you can see from Fig. 6.3b, the vertical positioning performance is significantly improved where estimates follow the real path very well. Fig. 6.4 shows the positioning errors cumulative distribution function (CDF) of the overall, horizontal and vertical coordinates that further confirms the accuracy of the proposed method. It should be noted that in this method after the horizontal coordinates are estimated with the presumed vertical coordinates, all the x, y and z coordinates are fed back into the algorithm for further modification. However, in the previous 3-D positioning algorithm explained in Chapter 4, only the z coordinate is fed back into the system to be updated.

The root mean square (RMS) errors and 95% confidence interval errors are also summarized in Table 6.1. The entire error is just 0.0252 m, and 95% of the errors are within 0.0432 m. Vertical RMS error is just 0.0120 m, and the 95% confidence interval error is 0.0199 m.

6.2 3-D POSITIONING BASED ON MULTIPLE RECEIVERS

We have thus far considered indoor positioning techniques using single receiver and multiple transmitters. A novel concept has been proposed in [3] for integrating visible

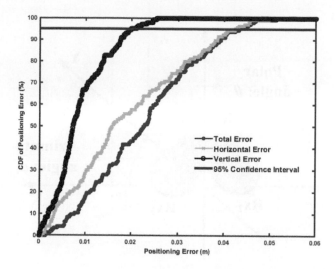

Figure 6.4: CDF curve of the 3-D positioning results.

light communication (VLC) with 3-D indoor positioning using a single transmitter and multiple tilted optical receivers. On the receiver side, three tilted optical receivers are gathered in a circle of 3 cm diameter. Based on the received signal strength (RSS) and AOA, 3-D positioning can be achieved.

The structure of receivers used in this method is shown in Fig. 6.5. As can be seen, there are three receivers, Rx_1, Rx_2 and Rx_3, with different orientations. These receivers have the same polar angle $10°$ but different azimuth angles as $90°$, $210°$ and $330°$, respectively.

The vertical view of the receiver is shown in Fig. 6.6. Three receivers which have a diameter of 1 cm are gathered in a circle of radius 1.5 cm. Also, the distance from the center of device to the center of receiver is 1 cm.

6.2.1 3-D POSITIONING ALGORITHMS

As mentioned in the previous chapters, RSS and AOA are the main positioning algorithms used in indoor VLC localization. In this section, we will first explain these methods in detail. Then, in contrast with conventional methods deploying single receiver and multiple transmitters for RSS and AOA algorithms, we employ multiple receivers and also combine these two methods to improve 3-D positioning.

6.2.1.1 Received Signal Strength Method

As discussed in Chapter 3, the main idea of RSS technique is using the attenuation of emitted signal strength to estimate the distance of the receiver from some set of transmitters. Since the positions of transmitters are fixed and known by the mobile

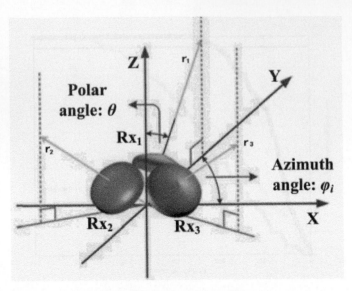

Figure 6.5: Structure of multiple tilted and separated optical receivers [4].

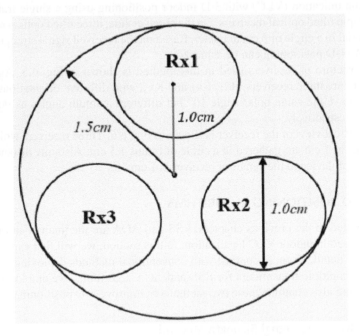

Figure 6.6: Vertical view of the receiver.

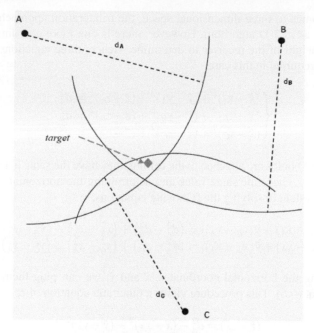

Figure 6.7: Principle of RSS.

device, it is able to obtain the location of receiver according to some geometric relationships and the estimated distances. Signal attenuation-based methods attempt to
calculate the distance from the transmitter to receiver according to signal path loss
in propagation.

In a two-dimensional space, the location of target can be determined by estimating
the distances from receiver to three reference transmitters [5]. As shown in Fig. 6.7,
the estimated horizontal distances from the receiver to transmitters are denoted by
d_A, d_B and d_C. There are three circles which have radius of d_A, d_B and d_C. Also, the
center of each circle is located at the transmitter or light emitting diode (LED) A,
B and C, correspondingly. Then, the target should be located at the intersection of
these three circles. Mathematically, the location of target, i.e., $(\tilde{x}, \tilde{y})$ can be obtained
by solving the following set of quadratic equations,

$$\begin{cases} (\tilde{x} - x_A)^2 + (\tilde{y} - y_A)^2 = d_A^2 \\ (\tilde{x} - x_B)^2 + (\tilde{y} - y_B)^2 = d_B^2 \\ (\tilde{x} - x_C)^2 + (\tilde{y} - y_C)^2 = d_C^2 \end{cases} \tag{6.5}$$

where (x_A, y_A), (x_B, y_B) and (x_C, y_C) are the horizontal coordinates of the fixed transmitters or LED bulbs. However, the set of equations in Eq. (6.5) may not have a feasible solution. Nevertheless, linear least square estimation (LLSE) can be applied to
obtain reliable results when the distances from receiver to transmitters are known [6].
By applying LLSE, the location of receiver can be estimated with at least three line-
of-sight (LoS) links.

When it comes to three-dimensional space, the trilateration approach is applied the same way as in 2-D algorithm. However, there is one more coordinate, i.e., $\tilde{z}$, which is the height of the receiver to determine. Hence, three equations as well as LoS links are required in this case.

$$\begin{cases} (\tilde{x} - x_A)^2 + (\tilde{y} - y_A)^2 + (\tilde{z} - z_A)^2 = d_A^2 \\ (\tilde{x} - x_B)^2 + (\tilde{y} - y_B)^2 + (\tilde{z} - z_B)^2 = d_B^2 \\ (\tilde{x} - x_C)^2 + (\tilde{y} - y_C)^2 + (\tilde{z} - z_C)^2 = d_C^2 \end{cases} \tag{6.6}$$

Since all LED bulbs are attached to the ceiling, they have the same heights. Therefore, z_A, z_B and z_C have the same value and determining the horizontal coordinates, $\tilde{x}$ and $\tilde{y}$ is equivalent to solving the following equations,

$$\begin{cases} 2\left[\tilde{x}(x_B - x_A) + \tilde{y}(y_B - y_A)\right] = \left(d_A^2 - d_B^2\right) + \left(x_B^2 - x_A^2\right) + \left(y_B^2 - y_A^2\right) \\ 2\left[\tilde{x}(x_C - x_A) + \tilde{y}(y_C - y_A)\right] = \left(d_A^2 - d_C^2\right) + \left(x_C^2 - x_A^2\right) + \left(y_C^2 - y_A^2\right) \end{cases} \tag{6.7}$$

After obtaining the horizontal coordinates, $\tilde{x}$ and $\tilde{y}$, we can plug them in the first equation of Eq. (6.6). This procedure yields a quadratic equation of $\tilde{z}$,

$$(\tilde{z} - z_A) = d_A^2 - (\tilde{x} - x_A)^2 - (\tilde{y} - y_A)^2 \tag{6.8}$$

As $\tilde{x}$ and $\tilde{y}$ are already known in Eq. (6.8), it is easy to obtain the value of $\tilde{z}$. Typically, there are two possible solutions of $\tilde{z}$ that one of them is larger than z_A and the other one is smaller than z_A. Since z_A is the height of ceiling, the height of receiver cannot be higher than z_A. Thus, we can simply pick the smaller solution giving us

$$\tilde{z} = z_A - \sqrt{d_A^2 - (\tilde{x} - x_A)^2 - (\tilde{y} - y_A)^2} \tag{6.9}$$

In [6–9], the receiving plane and transmitting plane are assumed to be parallel with each other. Based on this assumption, the irradiance angle ϕ and incidence angle ψ will have the same value. In this case, the distances from the receiver to LED bulbs d_A, d_B and d_C can be easily calculated according to the received optical power from different LED bulbs [7]. However, it is not guaranteed ϕ and ψ are equal. Hence, it is hard to calculate the distances d_A, d_B and d_C in general cases.

Let $\mathbf{n} = [n_x, n_y, n_z]$ denote the normal direction of the receiver. Since the heights of all LED bulbs are the same, we denote the difference of height between the receiver and transmitters as $h = z_A - \tilde{z}$. When LoS links exist for three transmitters, the variables $\tilde{x}$, $\tilde{y}$ and h satisfy the following set of quartic equations,

$$\begin{cases} Q_1 \left[(\tilde{x} - x_A)^2 + (\tilde{y} - y_A)^2 + h^2\right]^2 = n_x h(x_A - \tilde{x}) + n_y h(y_A - \tilde{y}) + n_z h^2 \\ Q_2 \left[(\tilde{x} - x_B)^2 + (\tilde{y} - y_B)^2 + h^2\right]^2 = n_x h(x_B - \tilde{x}) + n_y h(y_B - \tilde{y}) + n_z h^2 \\ Q_3 \left[(\tilde{x} - x_C)^2 + (\tilde{y} - y_C)^2 + h^2\right]^2 = n_x h(x_C - \tilde{x}) + n_y h(y_C - \tilde{y}) + n_z h^2 \end{cases} \tag{6.10}$$

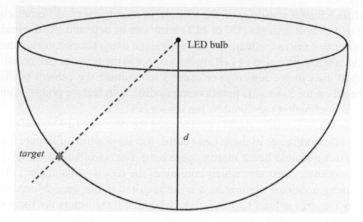

Figure 6.8: Principle of combined RSS and AoA.

In the equations above, Q_1, Q_2, Q_3 are constants given by

$$Q_i = \frac{\pi P_i}{T_s(\psi) g(\psi) P_t}, \quad i = 1, 2, 3 \tag{6.11}$$

where P_i is the received signal strength from the ith transmitter, P_t is the transmitted power, $T_s(\psi) = 1$ is the gain of optical filter, $g(\psi) = 1.5^2/\sin^2(70°)$ is the compound parabolic concentrator (CPC) gain at the receiver. Even though the equations in Eq. (6.10) can be solved numerically, the computational complexity of solving such high order equations is quite high.

Because visible light is not able to penetrate the walls or obstacles, there will be no interference among different rooms. In addition, the LoS component dominates the received optical power. Therefore, RSS is a practical and suitable approach in indoor positioning systems using visible light. However, the conventional RSS method used in [6, 7, 10] has some drawbacks. Since an RSS method requires at least three LoS links, the accuracy will be impaired significantly if the number of LoS links is less than three. Moreover, the coordinate of target is very difficult to obtain when the receiving and transmitting planes are not parallel. Therefore, the conventional RSS method may not be robust enough in general cases.

6.2.1.2 Angle of Arrival Method

Different from RSS, AOA belongs to angulation techniques. In AOA, the location of the desired target can be found by the intersection of several pairs of angle direction lines [11]. As shown in Fig. 6.8, AOA methods may use at least two known reference points, A and B, and two measured angles, θ_1 and θ_2 to derive the 2-D location of the target. Beamforming is applied to achieve AOA method in radio frequency (RF) [15]. As for visible light, image sensors are utilized to estimate the angle of arrival [12,13].

By utilizing image sensor(s) or the camera in smartphones, different colors of LED [12] or different patterns [13] of LED array can be detected. As the orientation of the receiver or camera inclination can be obtained using accelerometer and/or gyroscope, the relative direction of LED bulbs respect to the receiver can be estimated. AOA method uses image sensor(s) or camera to estimate the camera position and azimuth based on the 3D-to-2D point corresponding with feature points of an image as well as the inclination measured by the sensor [14].

Given camera inclination, the 2D positions of the detected LED bulbs on the image and the 3D coordinates of these LED bulbs, the 3D position of target $(\tilde{x}, \tilde{y}, \tilde{z})$ can be obtained using simple linear matrix operations. The set of solution $(\tilde{x}, \tilde{y}, \tilde{z})$ satisfies the orthonormal constraint which minimizes the reprojection errors [14]. Thus, time-consuming nonlinear optimization is not required in this case. Moreover, AOA method only requires at least two LoS links when the transmitters are located in the same horizontal plane [14].

Compared to RSS, AOA needs less LoS links and much lower computational costs. Since real-time performances are essential in indoor positioning systems, low computational complexity is an indispensable property. When the LED bulbs can be detected, AOA method using camera or image sensor(s) outperforms conventional RSS method using PD. Nevertheless, the resolution of image is sensitive to the motion of human body. Blurring images will enlarge the estimation error significantly. Besides, distance is another major limitation for AOA method. The accuracy of detecting LED bulbs is strongly affected by distance. To address the drawbacks of AOA and RSS methods, we will introduce a hybrid AOA and RSS method in the following.

6.2.1.3 Combined RSS and AoA Using Multiple Receivers

In Section 6.2.1.1, the basic concepts of RSS and AOA were introduced. In this subsection, we will develop an approach combining RSS and AOA and using multiple receivers. As for the receiver side, three tilted identical detectors (PDs) are deployed. In this approach, one single transmitter will be enough for the estimation of the receiver's location if the LoS link exists. By employing relationships among the received optical power of different detectors, the angle direction line between the receiver and LED bulb can be obtained. Then, we can easily get the incidence angle ψ and irradiance angle ϕ. At last, the distance between receiver and transmitter d can be derived according to Eqs. (2.6) and (2.7).

Now, we have determined the angle direction line and the distance d between the receiver and LED bulb. In a three-dimensional space, the set of points which has a distance d to the LED bulb will form a spherical surface centered at the LED bulb. Hence, the target should be located at the intersection of the angle direction line and the spherical surface as shown in Fig. 6.8.

There are three identical separate receivers (Rx_1, Rx_2 and Rx_3) at the receiver side. The orientations of the detectors in the three-dimensional space are denoted by three unit vectors as $\mathbf{r_1} = [x_1, y_1, z_1]$, $\mathbf{r_2} = [x_2, y_2, z_2]$ and $\mathbf{r_3} = [x_3, y_3, z_3]$. Also, $\hat{\mathbf{v}} = [\hat{x}, \hat{y}, \hat{z}]$ is the unit vector denoting the orientation of the angle direction line from the receiver to transmitter.

We can assume that the receivers in the device are located at the same place as they are quite close to each other. Under this assumption, the distance between the receiver and transmitter (d) and the irradiance angle (ϕ) are exactly the same for different receivers. In addition, the physical areas (A) of these detectors are also identical to each other. For the system under consideration, the concentrator gain $g(\psi)$ and the gain of optical filter $T_s(\psi)$ are constant. Therefore, the only different term in the direct current (DC) gain of different detectors is $\cos(\psi)$.

Given that LoS link exists for all three receivers and the Lambertian order $m = 1$, we can obtain

$$\begin{cases} x_1\hat{x} + y_1\hat{y} + z_1\hat{z} = \cos(\psi_1) \\ x_2\hat{x} + y_2\hat{y} + z_2\hat{z} = \cos(\psi_1)\frac{P_2}{P_1} \\ x_3\hat{x} + y_3\hat{y} + z_3\hat{z} = \cos(\psi_1)\frac{P_3}{P_1} \\ \hat{x}^2 + \hat{y}^2 + \hat{z}^2 = 1 \end{cases} \quad (6.12)$$

where P_1, P_2 and P_3 are the received optical power of direct light at receivers Rx_1, Rx_2 and Rx_3, respectively. If we express the first three equations above in the matrix form, we can obtain

$$\begin{bmatrix} x_1 & y_1 & z_1 \\ x_2 & y_2 & z_2 \\ x_3 & y_3 & z_3 \end{bmatrix} \begin{bmatrix} \hat{x} \\ \hat{y} \\ \hat{z} \end{bmatrix} = \cos(\psi_1) \begin{bmatrix} 1 \\ \frac{P_2}{P_1} \\ \frac{P_3}{P_1} \end{bmatrix} \quad (6.13)$$

In Eqs. (6.12) and (6.13), we actually use the concepts of dot product. Recall that the dot product of two vectors **a** and **b** is equal to $\|\mathbf{a}\| \, \|\mathbf{b}\| \cos(\theta)$, where $\|\mathbf{a}\|$ and $\|\mathbf{b}\|$ are the magnitudes of **a** and **b**, and θ is the angle between vector **a** and **b**.

Since the three vectors $\mathbf{r}_1 = [x_1, y_1, z_1]$, $\mathbf{r}_2 = [x_2, y_2, z_2]$ and $\mathbf{r}_3 = [x_3, y_3, z_3]$ are not in the same plane, the rows of 3-by-3 matrix are linear independent in Eq. (6.13). Because the FOV ψ_c is smaller than $90°$, and the incidence angle ψ is always smaller than FOV, the value of $\cos(\psi)$ is always larger than 0. Therefore, it is ensured that there exists unique $\hat{x}$, $\hat{y}$, $\hat{z}$ and $\cos(\psi_1)$ satisfying the constraint $\hat{x}^2 + \hat{y}^2 + \hat{z}^2 = 1$.

After obtaining the orientation of angle direction line $\hat{\mathbf{v}} = [\hat{x}, \hat{y}, \hat{z}]$, we can substitute it into the equation $x_1\hat{x} + y_1\hat{y} + z_1\hat{z} = \cos(\psi_1)$ and get the value of $\cos(\psi_1)$. Also, we can calculate the value of $\cos(\phi)$ easily as

$$\cos(\phi) = \sqrt{1 - \hat{z}^2}. \quad (6.14)$$

When the Lambertian order $m = 1$, the distance between receiver and transmitter d can be derived according to Eqs. (2.6) and (2.7) as

$$d = \sqrt{\frac{P_t}{\pi P_1} A \cos(\phi) T_s(\psi_1) g(\psi_1) \cos(\psi_1)}. \quad (6.15)$$

The location of target (x_R, y_R, z_R) can be then determined based on the distance d, the orientation of angle direction line $\hat{\mathbf{v}}$ and the position of LED bulb (x_T, y_T, z_T)

$$(x_R, y_R, z_R) = (x_T, y_T, z_T) - d\hat{\mathbf{v}} \quad (6.16)$$

In the previous derivations, we assume the positions of receivers as the center of the device. Nevertheless, this assumption will enhance the positioning error to some extent. To address this issue, we introduce a simple but effective iterative algorithm to adjust the estimated position. At first, we consider the position obtained in Eq. (6.16) as the initial position $(x_{(0)}, y_{(0)}, z_{(0)})$ for the center of device in the adjustment process. In the ith iteration, we calculate the real DC gains at current position $(x_{(i-1)}, y_{(i-1)}, z_{(i-1)})$ according to the Lambertian model, denoted by P_1', P_1' and P_1'. Also, we need to calculate the assumed DC gain $\bar{P}_k'$, $(k = 1, 2, 3)$ at the position $(x_{(i-1)}, y_{(i-1)}, z_{(i-1)})$ according to the measured DC gain P_1, P_2 and P_3. When the center of device is located at $(x_{(i-1)}, y_{(i-1)}, z_{(i-1)})$, we denote the real DC gain at the center of device as G_c, and the real DC gain at the kth $(k = 1, 2, 3)$ receiver as G_k. Then, we can obtain

$$\bar{P}_k = P_k \frac{G_c}{G_k} \quad k = 1, 2, 3. \tag{6.17}$$

Now, we can treat $\bar{P}_k$ as the measured DC gain of the kth receiver P_k and solve the linear equations in Eq. (6.13). In this way, we will get the updated orientation of angle direction line $[\hat{x}_{(i)}, \hat{y}_{(i)}, \hat{z}_{(i)}]$. This procedure aims to find the optimal angle direction line when the distance from the transmitter to the center of device is $d_{(i-1)}$. Then, we keep the distance $d_{(i-1)}$ and apply the updated angle direction line $[\hat{x}_{(i)}, \hat{y}_{(i)}, \hat{z}_{(i)}]$ to obtain a temporary position (x_s, y_s, z_s). The DC gain $\bar{P}_k'$ $(k = 1, 2, 3)$ at the position (x_s, y_s, z_s) is calculated again by Eq. (6.17). In addition, the real DC gain of the receivers P_k' $(k = 1, 2, 3)$ at the position (x_s, y_s, z_s) can be derived from the Lambertian model. In the next step, we try to find an appropriate distance $d_{(i)}$ minimizing the difference between $\bar{P}_k'$ and P_k'. When the angle direction line remains unchanged, the irradiance angle ψ and incidence angle ϕ do not change either. Under this circumstance, the DC gain P_k' is in proportion to $1/d^2$. In order to decrease the difference between P_k' and $\bar{P}_k'$, it is reasonable to make P_k' satisfy

$$\sum_{k=1}^{3} \bar{P}_k' = \sum_{k=1}^{3} P_k'. \tag{6.18}$$

Therefore, we can update the distance d as

$$d_{(i)} = d_{(i-1)} \sqrt{\frac{\sum_{k=1}^{3} \bar{P}_k'}{\sum_{k=1}^{3} P_k'}}. \tag{6.19}$$

The pseudocode for this iterative algorithm is shown in Table 6.2. In conclusion, each iteration can be decomposed into two steps. At first, find the optimal angle direction line at a fixed distance. Secondly, keep the angle direction line and find an optimal distance. If the distance between the point obtained in the $(i-1)^{th}$ iteration and the ith iteration is less than a small positive number ε or the number of iterations exceeds a specific integer, the adjustment algorithm will stop. If the algorithm doesn't converge at last, the estimated position will be regarded as an unreliable estimation.

Table 6.2

Pseudocode for adjusting the estimated position

Data

Measured DC gain of the receivers, P_1, P_2 and P_3

Initial position $(x(0), y(0), z(0))$

Initial orientation of angle direction line $\mathbf{r_0} = [\hat{x}(0), \hat{y}(0), \hat{z}(0)]$

Orientations of receivers $\mathbf{r_k} = [x_k, y_k, z_k]$ $k = 1, 2, 3$

Position of the transmitter (x_T, y_T, z_T)

Result:

The adjusted position of receiver (x_R, y_R, z_R)

Initialization;

i=0;

while $step > \varepsilon, i < maxIter$ **do**

1. Calculate the DC gain $\bar{P}_k$ ($k = 1, 2, 3$) at the position $(x(i-1), y(i-1), z(i-1))$ according to the measured DC gain P_k ($k = 1, 2, 3$), the real DC gain at the receiver G_k ($k = 1, 2, 3$) and the real DC gain at the center of device G_c;

2. Solve the following set of linear equations and obtain the updated angle direction line $[\hat{x}(i), \hat{y}(i), \hat{z}(i),]$

$$\begin{bmatrix} x_1 & y_1 & z_1 \\ x_2 & y_2 & z_2 \\ x_3 & y_3 & z_3 \end{bmatrix} \begin{bmatrix} \hat{x}_{(i)} \\ \hat{y}_{(i)} \\ \hat{z}_{(i)} \end{bmatrix} = \begin{bmatrix} \bar{P}_1 \\ \bar{P}_2 \\ \bar{P}_3 \end{bmatrix}$$

3. Obtain a temporary estimated position

$$(x_s, y_s, z_s) \leftarrow (x_T, y_T, z_T) - d_{(i-1)} \left(\hat{x}_{(i)}, \hat{y}_{(i)}, \hat{z}_{(i)} \right)$$

4. Calculate the supposed DC gain $\bar{P}'_k$ ($k = 1, 2, 3$) at the position $(x(i-1), y(i-1), z(i-1))$ according to the measured DC gain P_k ($k = 1, 2, 3$), the real DC gain at the receiver G_k ($k = 1, 2, 3$) and the real DC gain at the center of device G_c.

5. Calculate the real DC gain of the receivers at the position (x_s, y_s, z_s) according to Lambertian model, denoted by P'_k ($k = 1, 2, 3$)

6. Update the distance

$$d_{(i)} \leftarrow d_{(i-1)} \sqrt{\frac{\bar{P}'_1 + \bar{P}'_2 + \bar{P}'_3}{P'_1 + P'_2 + P'_3}}$$

7. Update the position

$$\left(x_{(i)}, y_{(i)}, z_{(i)} \right) \leftarrow (x_T, y_T, z_T) - d_{(i)} \left(\hat{x}_{(i)}, \hat{y}_{(i)}, \hat{z}_{(i)} \right)$$

8. $step \leftarrow \sqrt{ \left(x_{(i)} - x_{(i-1)} \right)^2 + \left(y_{(i)} - y_{(i-1)} \right)^2 + \left(z_{(i)} - z_{(i-1)} \right)^2 }$

9. $i \leftarrow i + 1$

end

return $\left(x_{(i)}, y_{(i)}, z_{(i)} \right)$

Table 6.3
Simulation results for the proposed positioning system

	Configuration I	Configuration II
Location of Transmitters	$(0.05, 0.05, 3.5)$ $(0.05, 5.95, 3.5)$ $(5.95, 0.05, 3.5)$ $(5.95, 5.95, 3.5)$	$(2, 2, 3.5)$ $(2, 4, 3.5)$ $(4, 2, 3.5)$ $(4, 4, 3.5)$
Mean Estimation Error Before Adjustment	0.1148 m	0.0882 m
Mean Estimation Error After Adjustment	0.0284 m	0.0080 m
Average Availability Rate	99.98%	93.65%

Previously, there are also some works applying combined RSS and AOA [3, 15] in the indoor positioning systems using visible light. In [3], the experiments have obtained an average error distance less than 3 cm and maximum error distance less than 6 cm in a 2 m × 2 m × 2.5 m room. However, the channel model observed in [3] doesn't accord with the Lambertian model. Instead, exponential model was more suitable in the experiments of [3]. Nevertheless, Lambertian model is suitable for VLC systems using LED [16–18].

In this section, we develop the combined RSS and AOA method where the channel is characterized by Lambertian model. In addition, only one transmitter is deployed in previous works [3]. In this case, the device will not be able to estimate the position if an arbitrary receiver does not have the LoS link. Therefore, we deploy four transmitters in the proposed system. With appropriate placement of transmitters, the probability that the device is not able to access LoS links will become very low.

In the combined RSS and AOA method, image sensors or camera are not used as receiver. Therefore, the motion of human will not bring any troubles to the detection of LED sources. When the LoS links exist, we can simply estimate the angle of arrival based on the power differences among different receivers. And the computational complexity is dominated by solving the set of linear equations in Eq. (6.10). Compared with conventional RSS method, the computational complexity is much lower in the combined RSS and AOA method even if the receiving plane is not parallel with the transmitting plane. Therefore, the combined RSS and AOA method outperforms conventional RSS and AOA method in both robustness and computational complexity.

6.2.2   SIMULATION AND RESULTS

In the computer simulation, the room size is 6 m × 6 m × 3.5 m, and there are totally four receivers which have two sets of locations. In terms of the first set, the coordinates of transmitter locations are $(0.05, 0.05, 3.5)$, $(0.05, 5.95, 3.5)$,

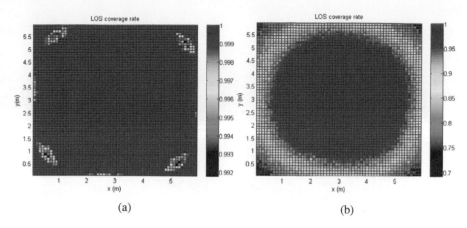

Figure 6.9: Availability rate of (a) Configuration I, and (b) Configuration II.

$(5.95, 0.05, 3.5)$ and $(5.95, 5.95, 3.5)$. In terms of the second set, the coordinates of transmitter locations are $(2, 2, 3.5)$, $(2, 4, 3.5)$, $(4, 2, 3.5)$ and $(4, 4, 3.5)$. Besides, we also assume that the orientations $(\mathbf{r_i} = [x_i, y_i, z_i], i = 1, 2, 3)$ of the receivers are measured by the gyroscope and/or accelerometers. Considering the height of human hands, the height of the device is generated randomly from 1.0 m to 1.4 m. When it comes to the horizontal locations, the coordinates of x-axis and y-axis are ranged from 0.05 m to 5.95 m where the distance between adjacent points is 0.1 m. Typically, the elevation angle is larger than 0 when people are looking at the screen of their mobile devices. Thus, the elevation angle of the device is randomly generated from $45°$ to $90°$. As for the azimuth angle of the device, it can be any value from $0°$ to $360°$. Hence, we generate the azimuth angle from a uniform random variable from $0°$. Moreover, the elevation angle of transmitters is $-90°$. Under this circumstance, the main simulation results are shown in Table 6.3.

The simulation is repeated for 5000 times, then the average availability rate and positioning error at each point in the horizontal plane are obtained. The device is able to estimate its position when all three receivers are able to get access to LoS links from any transmitter. If more than one transmitter is available, the device will estimate the position based on all available transmitters and apply the weighted average of all estimate positions, where the weight is the average DC gain from the corresponding transmitter.

At first, we will discuss the availability rate of the system. As for Configuration I, transmitters are located at the corners of the room and the average availability rate is very close to 100%. Shown in Fig. 6.9, the availability rate of Configuration I is 100% at most locations. The minimum available rate around 99.25% occurs at four locations $(0.55, 0.75)$, $(0.75, 5.55)$, $(5.15, 0.45)$ and $(0.45, 5.25)$. As for Configuration II, transmitters are located near the center of the room and the average availability

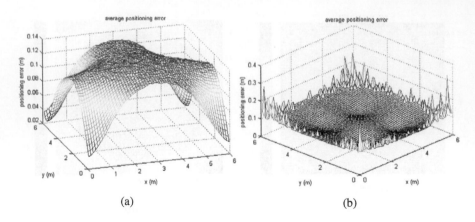

(a)   (b)

Figure 6.10: Positioning error before adjustment for (a) Configuration I, and (b) Configuration II.

rate comes down to 93.65%. The minimum available rate around 70.90% occurs at four corners of the room (0.05, 0.05), (0.05, 5.95), (5.95, 0.25) and (5.95, 5.95).

Then, we will talk about the positioning error in the proposed system. As it shows in Fig. 6.10, the average positioning error before adjustment for Configuration I is 0.1148 m and for Configuration II is 0.0882 m. In terms of the maximum positioning error, it is 0.1380 m for Configuration I, occurring at (1.35,3.65). Furthermore, the maximum positioning error before adjustment for Configuration II is 0.2880m, occurring at the corner (0.05, 0.05). After adjustment, the positioning error for Configuration I is reduced to 0.0284 m and the positioning error for Configuration I comes down to 0.0080 m. In addition, the maximum positioning error for Configuration II is 0.0746 m, which is much larger than the average value. Still, the maximum positioning error is yielded at the corner (0.05, 0.05) in Configuration II. In terms of Configuration I, the maximum positioning error is improved to 0.0498 m yielded at (2.95, 3.15) which is near the center of the room.

From Fig. 6.10 and Fig. 6.11, we can observe that the maximum positioning error is obtained when the distance between the transmitter and receiver approaches the maximum, and vice versa. In terms of Configuration I, the receivers are far from the transmitters when the device is located at the center of the room. But the receivers are more likely to get access to LoS links from different transmitters when the receivers are located at the center of the room. For Configuration II, no transmitter is near the corners. Also, it is hard to get reliable LoS links from different transmitters when the device is located at the corners. Therefore, both the positioning error and availability rate have poor performances in Configuration II as the device is located at the corners. Consequently, the maximum positioning error is higher in Configuration II.

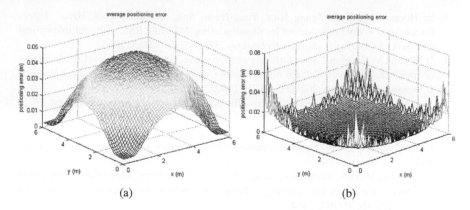

(a) (b)

Figure 6.11: Positioning error after adjustment for (a) Configuration I, and (b) Configuration II.

SUMMARY

In this chapter, a nonlinear model is constructed for the 3-D scenario. This technique better approximates the real situation when the height of the receiver is unknown. The trust region reflective algorithm acts as the solver.

Also in this chapter, a 3-D positioning algorithm combined RSS and AOA using multiple receivers were presented. This approach estimates the angle of arrival based on the power differences between different receivers. It is able to estimate the position when all three receivers are able to get access to LoS links from any transmitter. We tried two different configurations that have receivers placed at different locations in the simulation. The availability rates and average positioning error were obtained. It turns out that the available rates are higher than 90% for both configurations. When the transmitters are located at the corners, it achieves an availability rate of 99.98%.

In the proposed 3-D positioning algorithm, the positions of receivers are first presumed to be the center of device where unsatisfactory average positioning errors around 10 cm are obtained. In order to improve the positioning accuracy, an algorithm of position adjustment was developed. After adjusting, the estimated position iteratively, the average positioning error drops down to 2.84 cm and 0.80 cm for different configurations. Also, the maximum estimation error is 4.98 cm and 7.46 cm, respectively.

REFERENCES

1. Ya-xiang Yuan. A review of trust region algorithms for optimization. In *Iciam*, volume 99, pages 271–282, 2000.
2. Jorge J Moré. The Levenberg-Marquardt algorithm: implementation and theory. In *Numerical analysis*, pages 105–116. Springer, 1978.

3. Se-Hoon Yang, Hyun-Seung Kim, Yong-Hwan Son, and Sang-Kook Han. Three-dimensional visible light indoor localization using AOA and RSS with multiple optical receivers. *Journal of Lightwave Technology*, 32(14):2480–2485, 2014.

4. G Yun and M Kavehrad. Spot-diffusing and fly-eye receivers for indoor infrared wireless communications. In *1992 IEEE International Conference on Selected Topics in Wireless Communications*, pages 262–265. IEEE, 1992.

5. Mohsen Kavehrad. Sustainable energy-efficient wireless applications using light. *IEEE Communications Magazine*, 48(12):66–73, 2010.

6. W Zhang and M Kavehrad. A 2-d indoor localization system based on visible light led. In *2012 IEEE photonics society summer topical meeting series*, pages 80–81. IEEE, 2012.

7. Wenjun Gu, Weizhi Zhang, Mohsen Kavehrad, and Lihui Feng. Three-dimensional light positioning algorithm with filtering techniques for indoor environments. *Optical Engineering*, 53(10):107107, 2014.

8. S-H Yang, E-M Jeong, D-R Kim, H-S Kim, Y-H Son, and S-K Han. Indoor three-dimensional location estimation based on led visible light communication. *Electronics Letters*, 49(1):54–56, 2013.

9. Hyun-Seung Kim, Deok-Rae Kim, Se-Hoon Yang, Yong-Hwan Son, and Sang-Kook Han. Indoor positioning system based on carrier allocation visible light communication. In *Conference on Lasers and Electro-Optics/Pacific Rim*, page C327. Optical Society of America, 2011.

10. Liqun Li, Pan Hu, Chunyi Peng, Guobin Shen, and Feng Zhao. Epsilon: A visible light based positioning system. In *11th {USENIX} Symposium on Networked Systems Design and Implementation ({NSDI} 14)*, pages 331–343, 2014.

11. Hui Liu, Houshang Darabi, Pat Banerjee, and Jing Liu. Survey of wireless indoor positioning techniques and systems. *IEEE Transactions on Systems, Man, and Cybernetics, Part C (Applications and Reviews)*, 37(6):1067–1080, 2007.

12. Toshiya Tanaka and Shinichro Haruyama. New position detection method using image sensor and visible light leds. In *2009 Second International Conference on Machine Vision*, pages 150–153. IEEE, 2009.

13. Ye-Sheng Kuo, Pat Pannuto, Ko-Jen Hsiao, and Prabal Dutta. Luxapose: Indoor positioning with mobile phones and visible light. In *Proceedings of the 20th annual international conference on Mobile computing and networking*, pages 447–458. ACM, 2014.

14. Daisuke Kotake, Kiyohide Satoh, Shinji Uchiyama, and Hiroyuki Yamamoto. A hybrid and linear registration method utilizing inclination constraint. In *Proceedings of the 4th IEEE/ACM International Symposium on Mixed and Augmented Reality*, pages 140–149. IEEE Computer Society, 2005.

15. Gregary B Prince and Thomas DC Little. A two phase hybrid rss/aoa algorithm for indoor device localization using visible light. In *2012 IEEE Global Communications Conference (GLOBECOM)*, pages 3347–3352. IEEE, 2012.

16. Toshihiko Komine and Masao Nakagawa. Fundamental analysis for visible-light communication system using led lights. *IEEE transactions on Consumer Electronics*, 50(1):100–107, 2004.

17. Joseph M Kahn and John R Barry. Wireless infrared communications. *Proceedings of the IEEE*, 85(2):265–298, 1997.

18. Fritz R Gfeller and Urs Bapst. Wireless in-house data communication via diffuse infrared radiation. *Proceedings of the IEEE*, 67(11):1474–1486, 1979.

7 Impact of Multipath Reflections

In the previous chapters, all the research results are based on a line-of-sight (LoS) link, which is not feasible to track without added complexity in an indoors environment with millions of reflective rays from the surfaces of ceilings, walls, furniture or even shiny floors. Introducing the imaging (fly-eye) receivers in the latter parts of Chapter 6, this problem is alleviated due to the narrow field-of-view (FOV) of the individual eyes of the photodiodes (PDs).

In this chapter, for simplicity and economics of receiver design, we will again assume a single eye PD receiver per receiving terminal of the handset. Therefore, the multiple reflected rays cause severe multiple paths, and the receiver again will become dispersion-limited at higher transmit rates. This will severely limit the accuracy of the positioning operation and this is neglected in the majority of the reports that make a single received ray assumption.

As shown in Fig. 7.1, a multipath propagation phenomenon exists in the indoor wireless communications, i.e., the signals reach the receiver through more than one path. Multipath reflections degrade the communication quality as well as positioning accuracy. In this chapter, we will study the impact of multipath reflections on the positioning accuracy.

7.1 IMPULSE RESPONSE

Impulse response that usually characterizes the linear time-invariant (LTI) system, is used to describe the multipath reflections. When the input is a single, ideal Dirac pulse of electromagnetic power, its output is named as impulse response to describe the system with a function of time. Assume the transmitted signal is $x(t)$, with the knowledge of impulse response $h(t)$, the received signal can be expressed as:

$$y(t) = h(t) * x(t) \qquad (7.1)$$

Fig. 7.2 is an example of impulse response in frequency and time domains. Both experiment and simulation methods have been proposed to analyze the multipath reflections in the indoor optical wireless communication systems. Experimentally, the frequency response is measured first and then time domain response is obtained by applying inverse Fourier transform [1]. Simulation methods have been also proposed with different algorithms. In the following part, deterministic methods, modified Monte Carlo methods (MMC) and combined deterministic and combined deterministic and modified Monte Carlo (CDMMC) are explained.

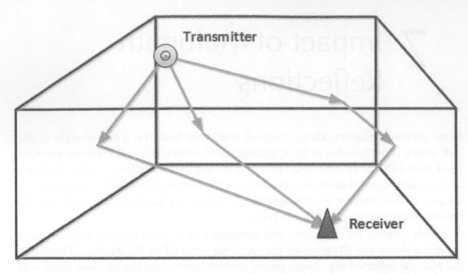

Figure 7.1: Geometry used to describe the multiple-reflection propagation.

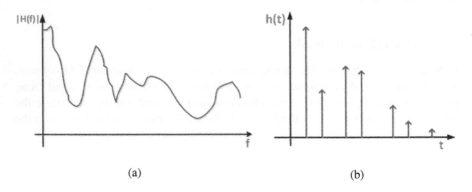

(a)   (b)

Figure 7.2: Example of impulse response in (a) frequency domain, and (b) time domain.

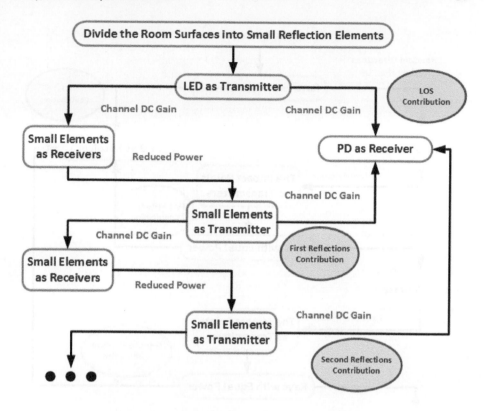

Figure 7.3: Flow diagram of Barry's approach.

7.2 DETERMINISTIC APPROACH

A deterministic approach has been proposed by Barry et al. where the surfaces of the entire room such as the walls, ceiling, floor, furniture are divided into small reflecting elements [2]. The LoS signal is calculated directly with the channel direct current (DC) gain as mentioned in Chapter 2. In order to calculate the first reflections, each reflecting element is considered as a receiver, and the received power from the LoS link is calculated.

The travelling times to the receivers are also recorded. These elements are then considered as separate transmitters, and their transmitted power is the received power degraded by the reflection coefficient. The power to the primary receiver, i.e., PD, from these elements through LoS link is calculated, and the travelling time is computed. The impulse response for the first reflections are calculated which includes the entire travelling time and received power.

To calculate the second reflections, each small reflector is considered as a receiver, and then all the other reflectors are considered as the transmitter. Similar to the calculation of first reflections, the received power of the PD is calculated when these small

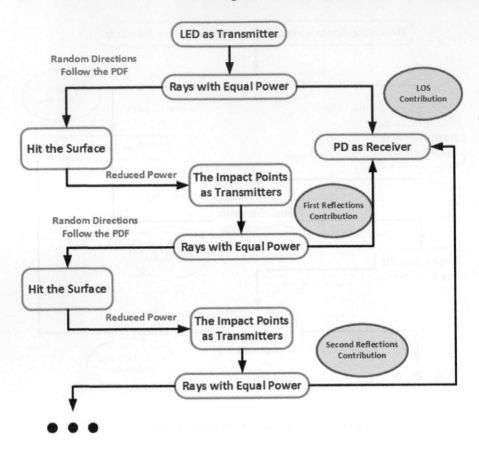

Figure 7.4: Flow diagram of the MMC method.

reflectors are considered as transmitters again. The entire travelling time is recorded. In this recursive way, subsequent contribution of reflections can be calculated. The smaller the reflecting elements are, the higher the approximation accuracy is. However, the computing time will increase considerably when the reflecting elements are smaller. Computing time increases tremendously after three orders of reflections with Barry's algorithm. Therefore, modification is needed to reduce the computational cost, especially in high-speed communication systems where higher order of reflections should be considered. The flow diagram of Barry's method is shown in Fig. 7.3.

7.3 MMC APPROACH

In order to decrease the computation time, an iterative approach named as Monte Carlo ray-tracking has been proposed [3]. In this method, the Lambertian pattern of the light emitting diode (LED) emission is treated as the probability density function (PDF). Rays are generated with equal optical power from the source and their directions follow the PDF. When each ray hits the surface of the room, the contact point is considered as a new source. New rays are generated from the new sources where their directions are generated randomly with the PDF and the power is reduced considering the reflectivity of the surface. The number of reflections is not limited as that of the deterministic method. However, the number of rays that finally reach the receiver is not large enough, which makes it necessary to generate a large number of rays resulting in a high computational cost.

As a solution, MMC approach has been proposed [4]. One ray to the receiver is generated from each source so that the number of rays from the source is reduced. This approach is fast but the calculated impulse response is not accurate enough considering variances are induced by the randomness of the directions of the rays. The variance can be reduced by increasing the number of rays, which in return brings up computation time. As MMC algorithm can be executed in parallel, the total computation time can be further reduced. The flow diagram of the MMC approach is shown in Fig. 7.4.

7.4 CDMMC METHOD

Barry's method is accurate while it suffers from extensive computation. On the other hand, the MMC algorithm speeds up the computation while the variances degrade the estimation accuracy. Therefore, a novel approach named CDMMC method, which has the advantages of the two proposed algorithms, has been proposed [5]. In this method, since the contribution of first reflections to the total impulse response is significant, they are calculated by Barry's method to ensure the accuracy. The contributions of the subsequent reflections are estimated by MMC method, where the computation time is decreased. Although MMC is not accurate enough, less power is remained after the first reflections so that the variance is acceptable. As there are many small reflection elements, the rays generated from each element do not need to be large in energy. This method is done as follows:

1. Divide the room surfaces into many small square elements, each of which has an area that is equal to the PD-receiving area.
2. The received power of the PD from the LED is calculated and the travelling time is recorded. The vector including power and time is considered as the LoS contribution to the total impulse response.
3. The small elements act as the receivers, and the received power is

$$P_{\text{received}}^{(0)} = H(0) P_{\text{source}}^{(0)} \tag{7.2}$$

where $P_{source}^{(0)}$ is the power emitted from the LED transmitter. In Eq. (7.2), $H(0)$ is the channel DC gain and $P_{\text{received}}^{(0)}$ is the received power of each small element. The travelling time is also tracked for each link.

4. Each of these small elements is considered as a point source again

$$P_{\text{source}}^{(1)} = \rho_{\text{surface}} P_{\text{received}}^{(0)} \qquad (7.3)$$

where ρ_{surface} is the reflection coefficient of the surfaces such as ceiling, floor and walls.

5. The received power of PD from each small element transmitter is calculated and the travelling time is recorded. These vectors including power and time are treated as the contribution of first reflections to the total impulse response.

6. MMC method is employed and N rays are generated from each small element transmitter sharing the equal power $P_{\text{ray}} = P_{\text{source}}^{(1)}/N$. The PDF of the rays' directions follows

$$f(\alpha, \beta) = \frac{m+1}{2\pi} \cos^m(\alpha) \qquad (7.4)$$

In Eq. (7.4), α is the angle between z-axis and the ray vector, β is the angle between projection of the ray vector on the X-Y plane and X-axis, and m is the Lambertian order. In Fig. 7.5, the origin point is each ray's point source and the X-Y plane represents the surface plane of the source. Note that Eq. (7.4) is independent of β. These rays hit the surface of the room with power $P_{\text{received}}^{(1)} = H(0) P_{\text{source}}^{(1)}$. The travelling time is tracked. The impact points are considered as new transmitters, where the power of these transmitters is

$$P_{\text{source}}^{(2)} = \frac{\rho_{\text{surface}} P_{\text{received}}^{(1)}}{N} \qquad (7.5)$$

One of the rays hits the PD, so the received power as well as the entire travelling time is recorded. In this way, the contributions of the second reflections are calculated. The other $N-1$ rays have the directions following the same PDF. Therefore, the subsequent reflections can be calculated iteratively.

Fig. 7.6 shows the flow diagram of the CDMMC method. The impulse response of the channel is computed by adding up all the contributions from LoS and each order of reflections [6, 7].

7.5 ANALYSIS OF IMPULSE RESPONSE

Consider that an entire room should be simulated including the corners and edges. We use a new model instead of just a cell. As shown in Fig. 7.7a, in our model, sixteen LEDs are installed on the ceiling of the room. Fig. 7.7b shows a bird's-eye view of the system model. The circles show the locations of all the LED bulbs, and the three selected locations marked with squares represent corner, edge and center. The area within the dashed line is the inside region, while the remaining area is considered as the outside region.

There are six reflection surfaces of the room, i.e., four walls, one ceiling and one floor, and they are assumed perpendicular to each other. The reflection coefficients are assumed to be fixed considering the material of the room surface and the 420 nm

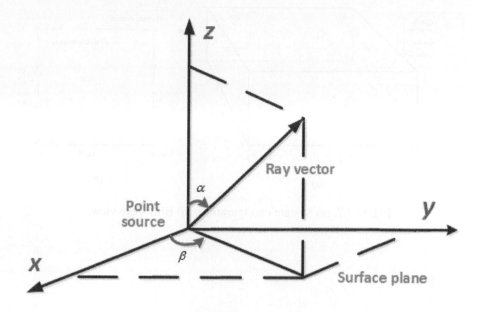

Figure 7.5: Coordinate system of a random ray.

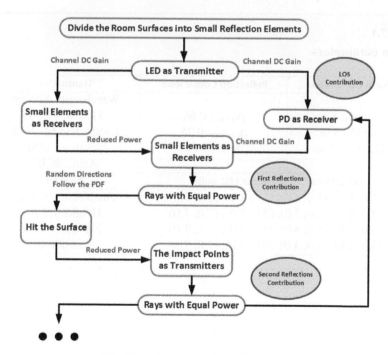

Figure 7.6: Flow diagram of the CDMMC method.

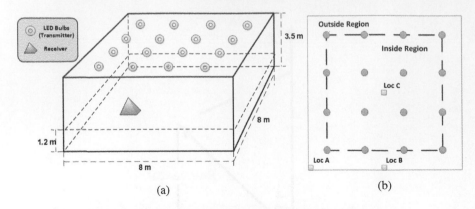

Figure 7.7: (a) System configuration. (b) Bird's-eye view.

Table 7.1
System parameters

Room Dimensions	Reflection Coefficients	Transmitter
Length: 8 m Width: 8 m Height: 3.5 m	ρ_{walls} : 0.66 ρ_{Ceiling} :0.35 ρ_{Floor} :0.60	Wavelength: 420 nm Height: 3.3 m Lambertian Mode: 1 Elevation: $-90°$ Azimuth: $0°$
Horizontal Coordinates of LED Bulbs		**Receiver**
(1.0, 1.0), (1.0, 3.0), (1.0, 5.0), (1.0, 7.0) (3.0, 1.0), (3.0, 3.0), (3.0, 5.0), (3.0, 7.0) (5.0, 1.0), (5.0, 3.0), (5.0, 5.0), (5.0, 7.0) (7.0, 1.0), (7.0, 3.0), (7.0, 5.0), (7.0, 7.0)		Area: $A = 10^{-4}\text{m}^2$ Height: 1.2 m Elevation: $90°$ Azimuth: $0°$

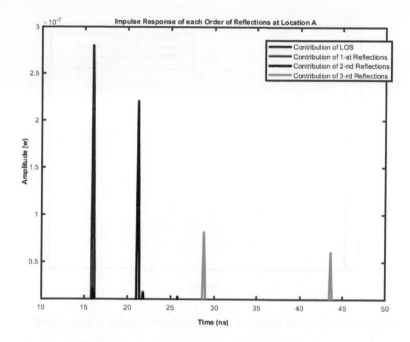

Figure 7.8: Impulse response of each reflection order at Location A.

light source. The transmitters are assumed as point sources and located at the height of 3.3 m considering the practical installation. The distance between the transmitters is assumed 2 m. As the transmitters are facing downwards, the azimuth angle is $0°$ and the elevation angle is $-90°$. The receiver is facing upwards. Thus, the azimuth angle is $0°$ and the elevation angle is $90°$. The receiving area of the PD is $10^{-4}\mathrm{m}^2$, with $70°$ FOV. The parameters of the model are summarized in Table 7.1.

Three typical locations are selected to analyze the effect of multipath reflections. Considering that the positioning system is low data rate, three orders of reflections are calculated. Location A with coordinates of (0 m, 0 m, 1.2 m) represents a point at the corner of the room, where the scatterings and reflections are severe. Location B with coordinates of (4 m, 0 m, 1.2 m) represents a point at the edge of the room, right beside the wall, where reflections are medium. Location C with coordinates of (4 m, 4 m, 1.2 m) represents central point where the effect of multipath reflections becomes weak. Considering the symmetrical property of the room, the impulse responses from the transmitter located at (3 m, 3 m, 3.3 m) are investigated at the three selected locations. The contribution of the LoS and the first three reflections are shown in Figs. 7.8 to 7.10.

Particularly, Fig. 7.8 demonstrates the impulse response of each reflection order at Location A. The impulse response amplitude of reflections is comparable to that

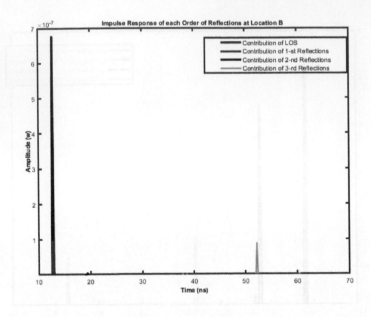

Figure 7.9: Impulse response of each reflection order at Location B.

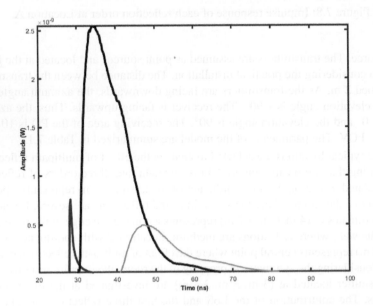

Figure 7.10: Impulse response of each reflection order at Location C.

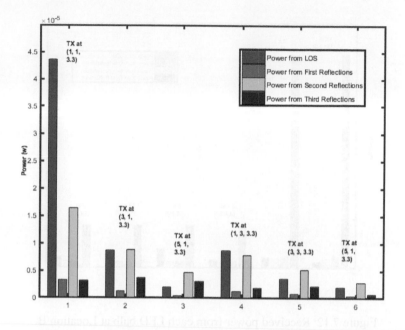

Figure 7.11: Received power from each LED bulb at Location A.

of LoS resulting in large positioning errors. Fig. 7.9 shows the impulse response of each reflection order at Location B. The amplitude of the reflections significantly decreases compared to Location A, and thus positioning accuracy is expected to be better than that at Location A. As shown in Fig. 7.10, at Location C, the LoS component almost dominates the total impulse response, and the amplitude of the reflections is negligible. Therefore, the positioning performance is expected to be less affected by multipath reflections.

7.6 POWER INTENSITY DISTRIBUTION ANALYSIS

As received signal strength (RSS) information is used to estimate the distance between a transmitter and receiver, the received power from each transmitter directly affects positioning performance. In this sub-section, we investigate received power from different LED transmitters at each reflection order for the three selected locations. Fig. 7.11 through Fig. 7.13 present the highest six received power values for the selected locations inside the room in descending order. The received power of each reflection order at the corner point is shown in Fig. 7.11. As can be seen clearly, only for the first LED, the LoS power value is much greater than that of the reflections. However, for the other five LED signals, the reflection components are

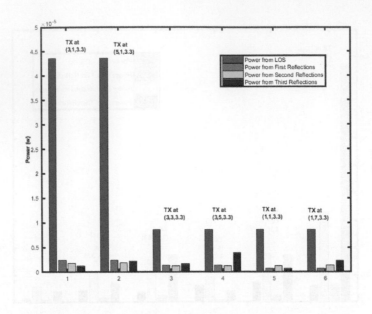

Figure 7.12: Received power from each LED bulb at Location B.

comparable to the LoS component. The reflection components affect positioning accuracy since only direct power attenuation from the transmitter is considered in the distance estimation.

Fig. 7.12 shows the received power of each reflection order at Location B. It is apparent that the received LoS power value is much greater than the received reflection power values from the first two LED transmitters. For the other four LED signals, the LoS power value significantly decreases, but it is still more than the reflection components. Therefore, the positioning error is expected to be smaller than that at Location A.

Fig. 7.13 shows the received power of each reflection order at Location C. For the first four strongest LED signals, the reflection components are negligible compared to the LoS component. Although for the other two LED signals, the LoS power value remarkably decreases, it is still much more than the reflection components. Therefore, the central point is expected to be less affected by multipath reflections.

7.7 POSITIONING ACCURACY

As a benchmark and in order to show the effect of multipath reflections on the positioning accuracy, positioning error neglecting the reflected power is also calculated and shown in Fig. 7.14. As can be seen, the positioning error is low all over the room, and only a little higher in the corner area. Fig. 7.15, on the other hand, shows the positioning performance considering the multipath reflections. It can be noted that each

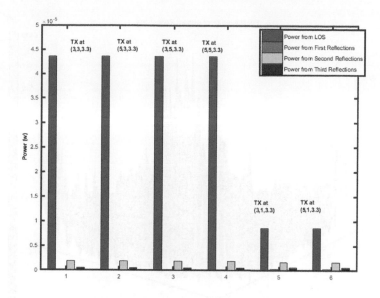

Figure 7.13: Received power from each LED bulb at Location C.

location of the room is affected by multipath reflections, especially the corner and edge area. However, the positioning accuracy is satisfactory at the central point of the room as reflections are weak there.

Fig. 7.16 presents the histograms of positioning errors when neglecting and considering multipath reflections, respectively. When no reflections are considered, the errors only come from the thermal noise and shot noise. In this ideal case, most of the errors are within 0.005 m. However, reflections cannot be practically avoided, and they are a major concern in the positioning system impairing dramatically the system performance as shown in Fig. 7.16b. In this case, most of the positioning errors are below 1 m while at some locations, the error climbs up to 1.7 m.

Fig. 7.17 shows the cumulative distribution function (CDF) of positioning errors of inner region, outer region and entire room. If only LoS link is considered, 95% of the errors are within 0.0085 m and the reflections raise the errors up to 1.41 m. Table 7.2 compares the positioning error quantitatively when neglecting and considering the reflections. At Location A, the error is 1.6544 m since the reflections are strong there. The effect of multipath reflections is medium at Location B, while the positioning performance is the best at Location C. The root mean square (RMS) error of the outside region is 0.8173 m due to severe reflections while the RMS error of the inside region is 0.2024 m. The RMS error of the entire room is 0.5589 m, while it is only 0.0040 m when no reflections are considered.

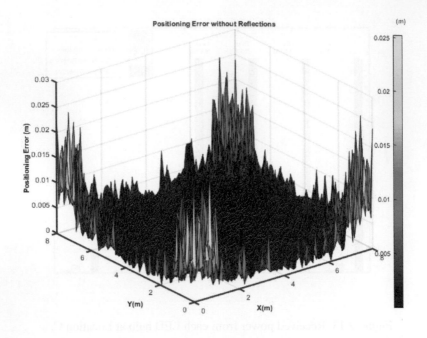

Figure 7.14: Positioning error considering no reflections.

Table 7.2

Positioning error neglecting/considering reflections

	Neglecting Reflections (m)	Considering Reflections (m)
Location A	0.0098	1.6544
Location B	0.0019	0.9966
Location C	0.0012	0.1674
Outside (RMS)	0.0059	0.8173
Outside (95% confidence interval)	0.0100	1.285
Inside (RMS)	0.0016	0.2024
Inside (95% confidence interval)	0.0032	0.382
Total (RMS)	0.0040	0.5589
Total (95% confidence interval)	0.0085	1.141

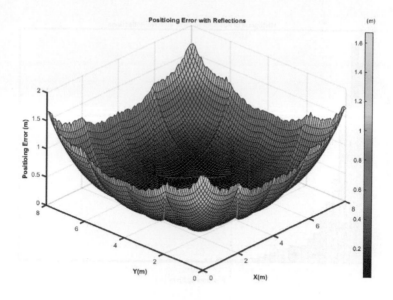

Figure 7.15: Positioning error considering reflections.

7.8 CALIBRATION APPROACHES

Since multipath reflections considerably affect the positioning accuracy, especially on the outer region, three calibration approaches are introduced here to improve the system performance.

7.8.1 NONLINEAR ESTIMATION

Nonlinear estimation can be used to improve the positioning accuracy when multipath reflections exist. Here, we consider a 2-D scenario where we first obtain $\hat{\mathbf{X}} = [\hat{x}, \hat{y}]^T$ from Eq. (4.15) as the initial value. In order to find an appropriate $\tilde{\mathbf{X}} = [\tilde{x}, \tilde{y}]^T$, the following equation should be minimized

$$\tilde{\mathbf{S}} = \sum_i \left(\sqrt{(x - x_i)^2 + (y - y_i)^2} - r_i \right)^2 \tag{7.6}$$

Trust region reflective algorithm is applied to estimate $\tilde{\mathbf{X}}$. With initial value $\tilde{\mathbf{X}}_0$, several points surrounding $\tilde{\mathbf{X}}_0$ are substituted to Eq. (7.6). The one minimizes $\tilde{\mathbf{S}}_1$ is selected as $\tilde{\mathbf{X}}_1$. After several iterative steps, receiver coordinates $\tilde{\mathbf{X}}$ will finally be obtained when $\tilde{\mathbf{S}}$ converges.

Fig. 7.18 demonstrates the positioning error distribution with the nonlinear estimation approach. At Location A, the error decreases from 1.6544 m with linear least square estimation (LLSE) to 1.1334 m with nonlinear estimation. At Location B,

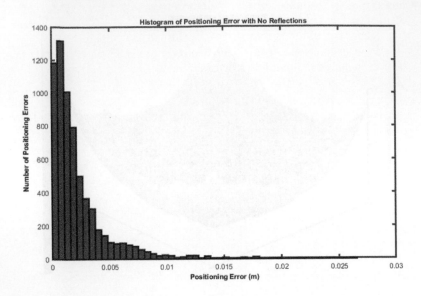

(a)

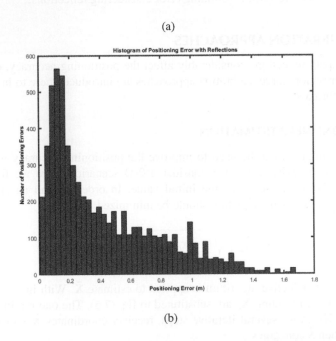

(b)

Figure 7.16: Histogram of positioning error (a) without reflections, and (b) with reflections.

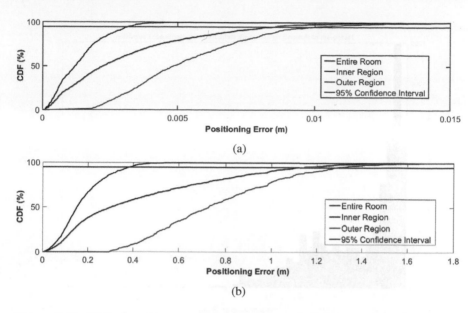

(a)

(b)

Figure 7.17: CDF of positioning errors (a) without reflections, and (b) with reflections.

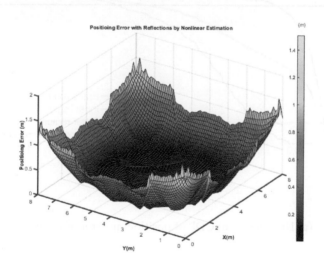

Figure 7.18: Positioning error with nonlinear estimation.

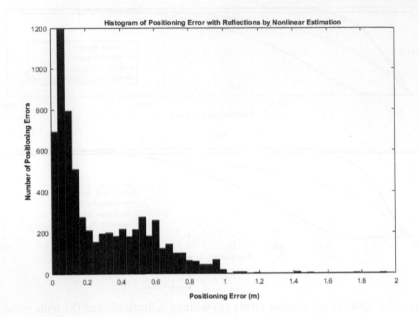

(a)

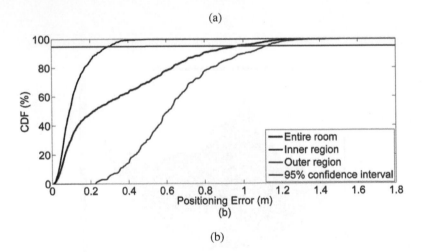

(b)

Figure 7.19: (a) Histogram. (b) CDF of positioning error of nonlinear estimation.

Table 7.3

Positioning error by nonlinear estimation (m)

	Inner (m)	Outer (m)	Total (m)
RMS	0.1401	0.6871	0.4642
95% Confidence Interval	0.2941	1.119	0.9681

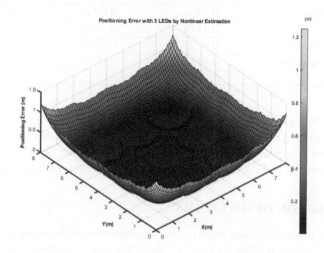

Figure 7.20: Positioning error with 3 LED signals by nonlinear estimation.

the error decreases from 0.9966 m to 0.8311 m. However, only slight improvement is seen at center point where the error decreases from 0.1674 m to 0.1427 m. The histogram of the positioning errors is shown in Fig. 7.19a. In this case, most of the errors are within 0.8 m, and only a few of them are over 1 m. The worst positioning error is just around 1.5 m. In Fig. 7.19b, the CDF distributions of inner region, outer region and entire room are shown respectively.

Table 7.3 shows the positioning performance of nonlinear estimation method in RMS and 95% confidence interval error. The nonlinear estimation outperforms original algorithm, especially for the outside region where the reflections are severe. In this region, the RMS error decreases from 0.8173 m to 0.6871 m. For the inner region, RMS error decreases from 0.2024 m to 0.1401 m and from 0.5589 m to 0.4642 m for the entire room. The 95% confidence interval error corresponding to inner region, outer and entire room also decreases, respectively.

In LLSE, we use Eq. (4.6) to approximate Eq. (4.5), but the mathematical deduction from Eq. (4.5) to Eq. (4.6) is not reversible. In other words, the optimum solution

Table 7.4

RMS error with transmitter selection approach

RMS error (m)	Outer Region	Inner Region	Entire Room
6 LEDs (Linear)	0.6760	0.1640	0.4616
6 LEDs (Nonlinear)	0.6016	0.1112	0.4046
5 LEDs (Linear)	0.5472	0.1251	0.3722
5 LEDs (Nonlinear)	0.5272	0.0851	0.3527
4 LEDs (Linear)	0.4838	0.0933	0.3259
4 LEDs (Nonlinear)	0.4828	0.0849	0.3240
3 LEDs (Linear)	0.4726	0.0917	0.3185
3 LEDs (Nonlinear)	0.4714	0.0863	0.3169

for Eq. (4.5) is applicable for Eq. (4.6), but the reverse is not always true. Therefore, previous estimation may induce approximation error. To avoid this error, the nonlinear estimation is applied by avoiding the approximation from Eq. (4.6) to Eq. (4.5), and solution of Eq. (4.5) is directly estimated through trust region algorithm.

7.8.2 SELECTION OF LED SIGNALS

The received power decreases when the distance between the transmitter and receiver increases. As shown in Figs. 7.11 to 7.13, the reflections contribute more to the total received power when the signal is from a further LED transmitter and brings larger errors in the distance estimation. Here, a signal selection approach is studied that the receiver only selects strong signals for the coordinate estimation. In our numerical analysis, the six, five, four and three strongest LED signals are selected, and RMS errors are calculated as shown in Table 7.4. By removing the signals affected considerably by multipath reflections, the positioning accuracy is improved. The total RMS error decreases to 0.4046 m, 0.3527 m, 0.3240 m and 0.3169 m, respectively, for the cases when the six, five, four and three strongest LED signals are selected. Note that, for the scenario with three LED bulbs, the strongest three that are not in a row must be selected to avoid singularity in matrix **A** of Eq. (4.7). As can be noted from Table 7.4, this approach improves the outer region accuracy more than the inner region.

For the sake of conciseness, only the best results are presented. Fig. 7.20 shows the positioning error distribution when the three strongest LED signals are selected for distance calculation and the nonlinear estimation is applied to obtain the receiver coordinates. Fig. 7.21b presents the corresponding histogram of positioning errors. It can be seen from Fig. 7.21a that many of the locations have positioning errors which are less than 0.4 m, and only a few locations have positioning error that is larger than 0.8 m. Fig. 7.21b is the CDF of the positioning errors where 95% confidence interval

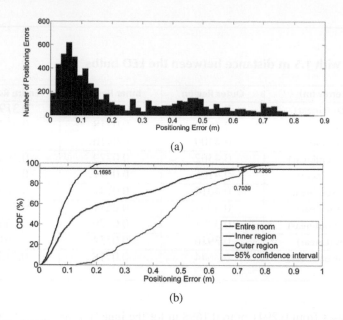

(a)

(b)

Figure 7.21: (a) Histogram. (b) CDF of positioning error of nonlinear estimation with 3 LED selected.

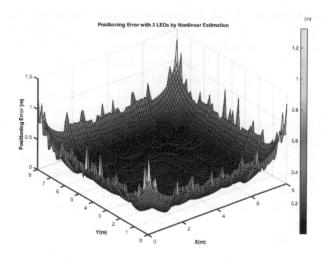

Figure 7.22: Positioning error with 3 LED signals by nonlinear estimation and 1.5 m distance between the LEDs.

Table 7.5

RMS error with 1.5 m distance between the LED bulbs

RMS error (m)	Outer Region	Inner Region	Entire Room
All LEDs (Linear)	0.4703	0.1052	0.3194
All LEDs (Nonlinear)	0.4672	0.0718	0.3121
6 LEDs (Linear)	0.4400	0.0916	0.2976
6 LEDs (Nonlinear)	0.4362	0.0729	0.2922
5 LEDs (Linear)	0.4106	0.0691	0.2751
5 LEDs (Nonlinear)	0.4014	0.0742	0.2699
4 LEDs (Linear)	0.4002	0.0571	0.2668
4 LEDs (Nonlinear)	0.3791	0.0729	0.2554
3 LEDs (Linear)	0.3916	0.0554	0.2610
3 LEDs (Nonlinear)	0.3644	0.0718	0.2458

error decreases from 0.2941 m to 0.1698 m for the inner region, from 0.9681 m to 0.7039 m for the outer region, and from 1.119 m to 0.7366 m for the entire room.

7.8.3 DECREASING THE DISTANCE BETWEEN LED BULBS

When LED bulbs are installed in a denser layout (i.e., the distance between LED bulbs is reduced, and a greater number of LED bulbs are used), the light intensity distribution becomes more uniform for the entire room, and therefore, the positioning accuracy is improved. Table 7.5 shows the RMS error where distance between the LED bulbs decreases to 1.5 m, and 25 LED bulbs are installed in total. With no LED signal selection, the entire RMS error is 0.3121 m with the nonlinear estimation. The RMS error decreases to 0.2922 m, 0.2699 m, 0.2554 m and 0.2458 m when six, five, four, and three LED signals are selected, respectively. The positioning errors are summarized in Table 7.5. The linear model, i.e., Eq. (7.6), is a simplified version of Eq. (7.5). Therefore, in most of the cases, the nonlinear estimation provides a better performance than the linear estimation. There are some cases that Eq. (7.6) approximates Eq. (7.5) much more accurately, and provides optimized solution. Meanwhile, the nonlinear estimation based on the trust region algorithm may not provide the optimized solution because of some convergence conditions. Therefore, the linear estimation outperforms its nonlinear counterpart in some cases of Table 7.5.

Fig. 7.22 shows the positioning performance for the best scenario, i.e., 3 LED signals selected for the nonlinear estimation. As can be seen from Fig. 7.22, for most of the inner area, the positioning performance is satisfactory while at the edges and corners of the room, there are some locations with large positioning errors. Fig. 7.23a is the histogram of the positioning errors where most of the errors are within 0.4 m, and there are only few outliers that can be removed with the filtering techniques

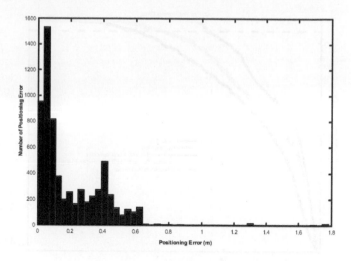

(a)

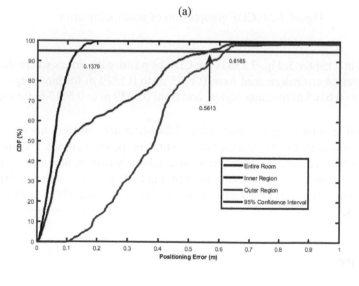

(b)

Figure 7.23: (a) Histogram. (b) CDF of positioning error with 3 LED signals by nonlinear estimation and 1.5 m distance between the LEDs.

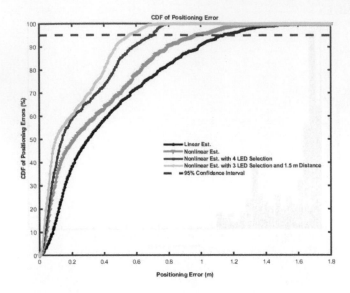

Figure 7.24: CDF comparison of positioning errors.

discussed in Chapter 5. Fig. 7.23b is the CDF of positioning errors where 95% confidence interval error decreased from 0.1695 m to 0.1379 m for inner region, from 0.7366 m to 0.6165 m for outer region, and from 0.7039 m to 0.5613 m for the entire room.

The CDF comparison presented in Fig. 7.24 better demonstrates improvement in the positioning performance and usefulness of the proposed methods. For linear estimation, the 95% confidence interval error is at 1.14 m, while the nonlinear estimation reduces it to 0.9681 m. Furthermore, by applying LED signal selection and decreasing the distance between LED bulbs, it can be improved to 0.7039 m and 0.5613 m, respectively.

SUMMARY

This chapter analyzes the impact of multipath reflections on the positioning accuracy in the complex indoor environment. The impulse response is analyzed by employing CDMMC algorithms. Impulse response and received signal power for three specific regions as corner, edge and center are characterized. The positioning error, histogram as well as CDF are plotted. To alleviate the influence of multipath reflections, three modification approaches are proposed. First, nonlinear estimation is introduced to better approximate the model. Second, signal selection is conducted so that the largely influenced signals can be removed. Finally, the distance between the LED bulbs is adjusted to make the light distributed more uniformly.

REFERENCES

1. Homayoun Hashemi, Gang Yun, M. Kavehrad, F. Behbahani, and Peter A. Galko. Indoor propagation measurements at infrared frequencies for wireless local area networks applications. *IEEE Transactions on Vehicular Technology*, 43(3):562–576, 1994.
2. John R. Barry, Joseph M. Kahn, William J. Krause, Edward A. Lee, and David G. Messerschmitt. Simulation of multipath impulse response for indoor wireless optical channels. *IEEE journal on selected areas in communications*, 11(3):367–379, 1993.
3. F.J. Lopez-Hernandez, R. Perez-Jimeniz, and A. Santamaria. Monte Carlo calculation of impulse response on diffuse IR wireless indoor channels. *Electronics Letters*, 34(12):1260–1262, 1998.
4. F.J. Lopez-Hernandez, R. Perez-Jimenez, and A. Santamaria. Modified Monte Carlo scheme for high-efficiency simulation of the impulse response on diffuse IR wireless indoor channels. *Electronics Letters*, 34(19):1819–1820, 1998.
5. M.I. Sakib Chowdhury, Weizhi Zhang, and Mohsen Kavehrad. Combined deterministic and modified Monte Carlo method for calculating impulse responses of indoor optical wireless channels. *Journal of Lightwave Technology*, 32(18):3132–3148, 2014.
6. Wenjun Gu, Mohammadreza Aminikashani, and Mohsen Kavehrad. Indoor visible light positioning system with multipath reflection analysis. In *2016 IEEE International Conference on Consumer Electronics (ICCE)*, pages 89–92. IEEE, 2016.
7. Wenjun Gu, Mohammadreza Aminikashani, Peng Deng, and Mohsen Kavehrad. Impact of multipath reflections on the performance of indoor visible light positioning systems. *Journal of Lightwave Technology*, 34(10):2578–2587, 2016.

8 OFDM Based Positioning Algorithm

Orthogonal frequency-division multiplexing (OFDM) has been applied to indoor wireless optical communications in order to mitigate the effect of multipath distortion of the optical channel as well as increasing data rate. In this chapter, an OFDM visible light communication (VLC) system is introduced which can be utilized for both communications and indoor positioning. We will demonstrate that the OFDM positioning system outperforms by 74% its conventional single carrier modulation scheme counterpart.

8.1 OFDM IN COMMUNICATION

OFDM is an effective modulation scheme by encoding the digital data on multiple orthogonal subcarriers. The OFDM technique has been widely applied in wireless communication systems since it performs well in reducing the intersymbol-interference (ISI) when data rate is high [1]. As shown in Fig. 8.1, compared with the traditional frequency-division multiplexing (FDM) scheme, the subcarrier signals used by OFDM are orthogonal with each other to save the bandwidth. The modulation of these subcarriers follows the traditional schemes such as quadrature amplitude modulation (QAM) and phase-shift keying (PSK). As each subcarrier is transmitted in low data rate, equalization process is simplified and the channel requirement is decreased.

There are several advantages of OFDM. First of all, OFDM can achieve high spectral efficiency. Second, it is robust against narrow-band and severe channel conditions contributing to less requirement on the equalization process. Third, it performs well with the help of guard intervals (GIs) when there is severe fading caused by multipath propagation and ISI. However, OFDM also suffers from several shortcomings such as being sensitive to Doppler shift and frequency synchronization and high peak-to-average-power ratio (PAPR). The cyclic prefix, GI and even training sequence also result in low transmission efficiency.

The orthogonality of OFDM signal is usually implemented with fast Fourier transform (FFT) algorithm. Inverse fast Fourier transform (IFFT) is applied on the sender side to generate orthogonal subcarriers and FFT is applied on the receiver side for the demodulation. GI is inserted between OFDM symbols to alleviate the ISI caused by multipath reflections, eliminate the requirement on the pulse-shaping filter and reduce the sensitivity to time synchronization problems.

Considering so many advantages of OFDM, it is widely proposed in various wireless communication standards such as digital audio broadcasting, digital video broadcasting, wireless local area network (WLAN) (802.11, HIPERLAN), and so forth.

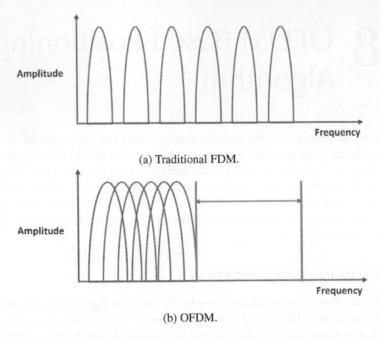

(a) Traditional FDM.

(b) OFDM.

Figure 8.1: OFDM modulation scheme compared with traditional FDM.

8.2 OFDM FOR VLC

Since the complex indoor environments commonly have many obstacles, multipath reflections are considered as one of the major factors degrading the communication performance. OFDM is proposed as a modulation scheme to combat the ISI incurred by the multipath reflections. There are also specific drawbacks in the OFDM based VLC system. High PAPR requires a wide dynamic range for the linear power amplifier. As a result, some nonlinear characteristics of the light emitting diode (LED) transmitter will largely impair the communication quality.

Since intensity modulation (IM)/direct detection (DD) is used in the VLC system, the transmitted signal must be positive and real. Therefore, Hermitian symmetry is always required for the input data to ensure the data to be real. In addition, several methods have been proposed to ensure the positivity including pulse-amplitude-modulated discrete multitoned (PAM-DMT) OFDM, DC-clipped optical OFDM (DCO-OFDM) as well as asymmetrically clipped optical OFDM (ACO-OFDM). PAM-DMT only modulates the imaginary parts of the subcarriers, and then the entire negative parts of the waveform are clipped off. The clipping noise only falls on the real part of each subcarrier and is orthogonal to the desired signal [2]. DCO-OFDM works by adding a direct current (DC) bias to the signal and then a hard clipping is carried out on the negative signal pulses. In ACO-OFDM, only odd subcarriers are modulated and the impairment from clipping noise is avoided [3,4].

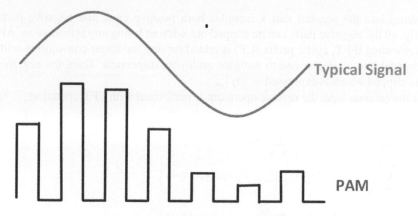

Figure 8.2: PAM constellation.

8.2.1 PAM-DMT

Let us assume N complex symbols as $\mathbf{I} = [I_0, I_1, ... I_{N-1}]$ represent the input bits, where $[.]^T$ denotes the transpose of a vector. Considering the requirement of Hermitian symmetry, the modulated symbol is written as $\tilde{\mathbf{I}} = [0, I_0, I_1, ... I_{N-1}, 0, I_{N-1}^*, ... I_1^*, I_0^*]$. I_m is represented as ja_m where a_m is real symbol mapped from a PAM constellation as shown in Fig. 8.2.

After conducting IFFT operation, the output signal can be obtained as

$$x_k = \frac{1}{2N} \sum_{m=0}^{2N-1} I_m e^{j2\pi \frac{m}{2N} k}$$

$$= \frac{1}{2N} \left[\sum_{m=0}^{N-1} I_m e^{j2\pi \frac{m}{2N} k} + \sum_{m=0}^{N-1} I_{2N-m} e^{j2\pi \frac{(2N-m)}{2N} k} \right]$$

$$= \frac{1}{2N} \sum_{m=0}^{N-1} i \left(a_m e^{j2\pi \frac{m}{2N} k} - a_m e^{-j2\pi \frac{m}{2N} k} \right) \tag{8.1}$$

$$= \frac{-1}{N} \sum_{m=0}^{N-1} a_m \sin\left(2\pi \frac{m}{2N} k \right) \qquad k = 0, 1, ..., 2N-1$$

Therefore, the symbols have the following property

$$x_{2N-k} = \frac{-1}{N} \sum_{m=0}^{N-1} a_m \sin\left(2\pi \frac{m}{2N} (2N - k) \right)$$

$$= \frac{1}{N} \sum_{m=0}^{N-1} a_m \sin\left(2\pi \frac{m}{2N} k \right) = -x_k \tag{8.2}$$

The entire symbol can be then written as:

$$\mathbf{x} = [0, x_0, x_1, ..., x_{N-1}, 0, -x_{N-1}, ..., -x_1, -x_0] \tag{8.3}$$

Taking into the account that **x** includes both positive parts and negative parts equally, all the negative parts can be clipped out without losing any information. After performing IFFT, cyclic prefix (CP) is added turning the linear convolution with the channel into a circular one to mitigate multipath dispersion. Then, the negative part is clipped which is expressed as $\lfloor x_k \rfloor_c$.

At the receiver side, the reverse operation is performed with FFT operation

$$
\begin{aligned}
\hat{I}_m &= \sum_{k=0}^{N-1} \lfloor x_k \rfloor_c e^{-j2\pi \frac{m}{2N} k} + \sum_{k=0}^{N-1} \lfloor x_{2N-k} \rfloor_c e^{-j2\pi \frac{m}{2N}(2N-k)} \\
&= \sum_{k=0}^{N-1} \left\{ \lfloor x_k \rfloor_c e^{-j2\pi \frac{m}{2N} k} + \lfloor -x_k \rfloor_c e^{-j2\pi \frac{m}{2N}(2N-k)} \right\} \\
&= \sum_{k=0}^{N-1} \left\{ \lfloor x_k \rfloor_c \left[\cos\left(2\pi \frac{m}{2N} k \right) - i \sin\left(2\pi \frac{m}{2N} k \right) \right] \right. \\
&\quad \left. + \lfloor -x_k \rfloor_c \left[\cos\left(2\pi \frac{m}{2N} k \right) + i \sin\left(2\pi \frac{m}{2N} k \right) \right] \right\} \\
&= \sum_{k=0}^{N-1} \left\{ [\lfloor x_k \rfloor_c + \lfloor -x_k \rfloor_c] \cos\left(2\pi \frac{m}{2N} k \right) - i [\lfloor x_k \rfloor_c - \lfloor -x_k \rfloor_c] \sin\left(2\pi \frac{m}{2N} k \right) \right\} \\
&= \sum_{k=0}^{N-1} \left\{ |x_k| \cos\left(2\pi \frac{m}{2N} k \right) - i [x_k] \sin\left(2\pi \frac{m}{2N} k \right) \right\}.
\end{aligned}
\tag{8.4}
$$

As shown in Eq. (8.4), the clipping noise mainly influence the real part of the subcarriers, and the symbols can be recovered with the imaginary part of the first half subcarriers.

8.2.2 DCO-OFDM

DCO-OFDM works in a straightforward way where a DC bias is added to the symbols to make them positive. The same symbol $\tilde{\mathbf{I}} = \left[0, I_0, I_1, \ldots I_{N-1}, 0, I_{N-1}^*, \ldots I_1^*, I_0^* \right]$ is generated but not constrained to PAM modulation. After IFFT operation, **x** becomes real but not always positive

$$
x_k = \frac{1}{2N} \sum_{m=0}^{2N-1} I_m \exp\left(\frac{j2\pi km}{2N} \right).
\tag{8.5}
$$

After the parallel to serial (P/S) converter, CP is added and then the analog signal $x(t)$ passes through a low-pass filter. An appropriate DC bias given by

$$
B_{DC} = \mu \sqrt{E\left\{ x(t)^2 \right\}}
\tag{8.6}
$$

is then added to the signal where μ is a proportionality constant and $x(t)$ is assumed to follow a Gaussian model with a zero mean and a variance $E\left\{ x_k^2 \right\}$. Since OFDM

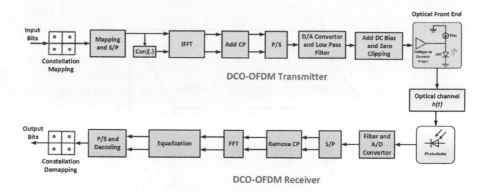

Figure 8.3: General block diagram of DCO-OFDM.

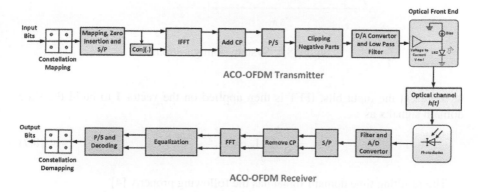

Figure 8.4: General block diagram of ACO-OFDM.

signals have a high PAPR, negative peaks still exist, and the clipping process induces noise. This noise can be reduced by a relatively larger bias, which in turn increases the optical energy/bit to noise power spectral density ratio E_b/N_0. As a result, DCO-OFDM scheme has the disadvantage of low optical power efficiency. The flow diagram of DCO-OFDM is shown in Fig. 8.3.

8.2.3 ACO-OFDM

In ACO-OFDM scheme, only odd subcarriers are modulated which results in avoiding the impairment from clipping noise. The block diagram of an ACO-OFDM system is depicted in Fig. 8.4.

In ACO-OFDM technique, during T sec, $N/4$ complex symbols are transmitted after a M-QAM mapping where $M = 4, 8, 16$, etc. To avoid the clipping noise and ensure a real output signal used to modulate the LED intensity, a set of N complex data symbols as $\mathbf{I} = \left[0, I_0, 0, I_1, ..., 0, I_{N/4-1}, 0, I_{N/4-1}^*, 0, ..., I_1^*, 0, I_0^*\right]^T$ are used

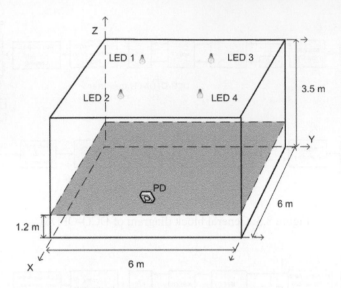

Figure 8.5: System configuration.

to represent the input bits. IFFT is then applied on the vector **I** to build the time domain signal **x** as

$$x_k = \frac{1}{N} \sum_{m=0}^{N-1} I_m \exp\left(\frac{j2\pi km}{N}\right). \tag{8.7}$$

The resulting time domain signal has the following property [4]

$$x_k = -x_{N/2+k} \quad k = 0, 1, \ldots, N/2 - 1 \tag{8.8}$$

To make the transmitted signal unipolar, all the negative values are clipped to zero. It is proven that since only the odd subcarriers are used to carry the data symbols, the clipping does not affect the data-carrying subcarriers, but only reduces their amplitude by a factor of two. The unipolar signal, $\lfloor x_k \rfloor_c$, is then converted to analog and filtered to modulate the intensity of an LED. At the receiver, the signal is converted back to digital. CP is then removed and the electrical OFDM signal is demodulated by taking a N point FFT and equalized with a single-tap equalizer on each subcarrier to compensate for channel distortion. The even subcarriers are then discarded and the transmitted data is recovered by a hard or soft decision.

8.3   OFDM IN VLC POSITIONING

In this section, an OFDM VLC system is studied which can be utilized for both communications and indoor positioning where a positioning algorithm based on power attenuation is used to estimate the receiver coordinates.

Table 8.1

System configuration

Room dimensions	Reflection coefficients
Length: 6 m Width: 6 m Height: 3.5 m	ρ_{wall}: 0.66 $\rho_{Ceiling}$: 0.35 ρ_{Floor}: 0.60
Transmitters (Sources)	**Receiver**
Wavelength: 420 nm Height(H): 3.3 m Lambertian mode (m): 1 Elevation: $-90°$ Azimuth: $0°$ Coordinates: (2,2) (2,4) (4,2) (4,4) Power for "1"/ "0": 5 W/3 W	Area (A_r): $1 \times 10^{-4} \, \text{m}^2$ Height (h): 1.2 m Elevation: $90°$ Azimuth: $0°$ FOV: $70°$

8.3.1 SYSTEM MODEL

We consider a typical room shown in Fig. 8.5 with dimensions of 6 m × 6 m × 3.5 m where four LED bulbs are located at a height of 3.3 m with a rectangular layout. Data are transmitted from these LED bulbs after driver circuits modulate them. Each LED bulb has an identification (ID) denoting its coordinates which is included in the transmitted data. A photodiode (PD) as the receiver is located at the height of 1.2 m and has a field-of-view (FOV) of 70° and a receiving area of 1 squared centimeter. The room configuration is summarized in Table 8.1. Furthermore, strict time domain multiplexing is used where the entire OFDM frequency spectrum is assigned to a single LED transmitter for at least one OFDM symbol including a CP.

For the sake of brevity, ACO-OFDM is considered in this chapter, as it utilizes a large dynamic range of LED and thus is more efficient in terms of optical power than systems using DC biasing. However, the generalization to other techniques is very straightforward.

A block diagram of an ACO-OFDM communication and positioning system is depicted in Fig. 8.6. The LED-IDs are encoded as the input bits and mapped to the M-QAM constellation as $\mathbf{I} = [I_0, I_1, ...I_{N-1}]$, the real and non-negative signals are generated to modulate the light intensity.

These modulated signals passing through the optical channel, taking the multipath reflections as well as shot noise and thermal noise into consideration. The channel impulse response is estimated by combined deterministic and modified Monte Carlo (CDMMC) method, where the line-of-sight (LoS) and first three reflections are considered to simulate the impulse response of the channel. For each transmitter, 4096 different channels are generated by placing the receiver in different locations within the room with the same height (i.e., 1.2 m). Figs. 8.7 to 8.9 demonstrate the

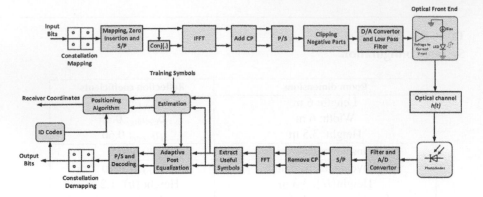

Figure 8.6: OFDM transmitter and receiver configuration for both positioning and communication purposes.

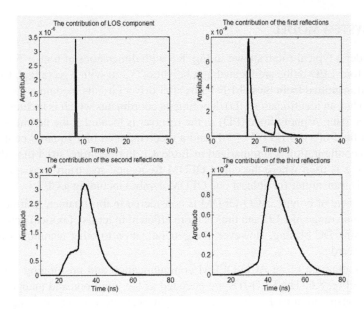

Figure 8.7: The contributions from different orders of reflections to the total impulse response of a location at the center of the room (weak scatterings and multipath reflections).

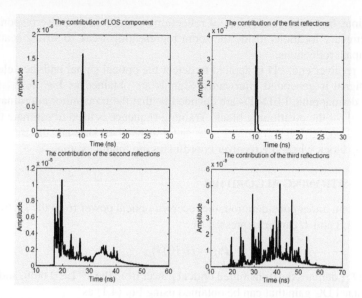

Figure 8.8: The contributions from different orders of reflections to the total impulse response of a location at the edge of the room (medium scatterings and multipath reflections).

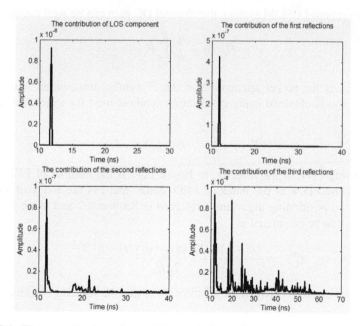

Figure 8.9: The contributions from different orders of reflections to the total impulse response of a location at the corner of the room (strong scatterings and multipath reflections).

contributions from different orders of reflections to the total impulse responses for three exemplary locations inside the room representing weak to strong scatterings and multipath reflections.

At the receiver end, PD is applied to detect the optical signal and then electrical domain signal is generated afterwards. Signals are obtained as $\hat{\mathbf{I}} = [\hat{I}_0, \hat{I}_1, ...\hat{I}_{N-1}]$ and after de-mapping, LED-IDs are decoded so that the transmitter coordinates can be obtained for the positioning block. Training sequence is used to estimate the signal attenuation, so $\mathbf{I}$ at transmitter side and $\hat{\mathbf{I}}$ at receiver side are the input of the positioning block where the receiver coordinates are finally estimated.

8.3.2 POSITIONING ALGORITHM

For the system under consideration, the received optical power from the kth transmitter, $k = 1, 2, 3$ and 4, can be expressed as

$$P_{r,k} = H_k(0) P_{t,k} \qquad (8.9)$$

where $P_{t,k}$ denotes the transmitted optical power from the kth LED bulb, and $H_k(0)$ is the channel DC gain that can be obtained using Eq. (4.1) as

$$H_k(0) = \frac{m+1}{2\pi d_k^2} A_r \cos^m(\phi_k) T_s(\psi_k) g(\psi_k) \cos(\psi_k) \qquad (8.10)$$

For the proposed OFDM system, the channel DC gain can be well estimated as[1]

$$\tilde{H}_k(0) = \frac{1}{N} \sum_{i=1}^{N} P_{k,i} \qquad (8.11)$$

where $P_{k,i}$ is the power attenuation of the ith symbol transmitted from the kth transmitter and is obtained using the training symbols used for synchronization as

$$P_{k,i} = \left| \frac{\hat{I}_i^k}{I_i^k} \right|^2 \qquad k = 1, 2, \dots, l \qquad (8.12)$$

Considering FOV of the receiver, in Eq. (8.12) l is the number of LED signals received by the PD, k is the index of LED signal, and i is the index of subcarriers. Similar to positioning algorithm explained in Section 4.2 and using Eqs. (8.9) to (8.12), d_k can be calculated as

$$d_k^{m+3} = \frac{(m+1) A_r T_s(\psi_k) g(\psi_k) (H-h)^{m+1}}{2\pi \tilde{H}_k}. \qquad (8.13)$$

Horizontal distance between the kth transmitter and the receiver can be estimated as

$$r_k = \sqrt{d_k^2 - (H-h)^2}. \qquad (8.14)$$

---

[1] Note that for the other OFDM techniques, it is only required to estimate the channel DC gain similarly using the sample mean of the power attenuation of the transmitted symbols. Therefore, this system can be easily utilized for other OFDM techniques as well.

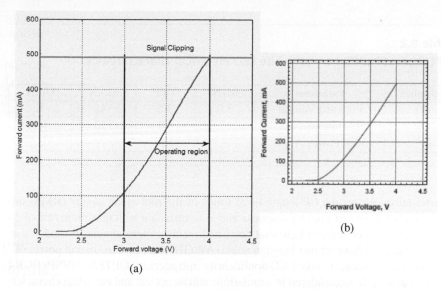

Figure 8.10: Transfer characteristics of OPTEK, OVSPxBCR4 1-Watt white LED. (a) Fifth-order polynomial fit to the data. (b) The curve from the data sheet.

Then, according to the lateration algorithm discussed in more detail in Chapter 4, the estimated receiver coordinates can then be obtained by the linear least squares estimation approach as

$$\hat{\mathbf{X}} = (\mathbf{A}^T \mathbf{A})^{-1} \mathbf{A}^T \mathbf{B} \tag{8.15}$$

where $\mathbf{A}$ and $\mathbf{B}$ are defined as

$$\mathbf{A} = \begin{bmatrix} x_2 - x_1 & y_2 - y_1 \\ x_3 - x_1 & y_3 - y_1 \\ x_4 - x_1 & y_4 - y_1 \end{bmatrix} \tag{8.16}$$

$$\mathbf{B} = \frac{1}{2} \begin{bmatrix} \left(r_1^2 - r_2^2\right) + \left(x_2^2 + y_2^2\right) - \left(x_1^2 + y_1^2\right) \\ \left(r_1^2 - r_3^2\right) + \left(x_3^2 + y_3^2\right) - \left(x_1^2 + y_1^2\right) \\ \left(r_1^2 - r_4^2\right) + \left(x_4^2 + y_4^2\right) - \left(x_1^2 + y_1^2\right) \end{bmatrix} \tag{8.17}$$

In Eqs. (8.15) to (8.17), $\mathbf{X} = [x_c, y_c]^T$ is the receiver coordinates to be estimated and (x_k, y_k) is the kth transmitter coordinates obtained from the recovered LED ID in a two-dimensional space.

8.4   SIMULATION AND ANALYSIS

In this section, we present numerical results for the proposed indoor VLC system. In the following, an OFDM system with a number of subcarriers of $L = 64, 256, 512$ or 1024 is considered where the symbols are drawn from an M-QAM modulation

Table 8.2
OPTEK, OVSPxBCR4 1-watt white LED electrical characteristics

Symbol	Parameter	MIN	TYP	MAX	Units
V_F	Forward Voltage	3.0	3.5	4	V
Φ	Luminous Flux	67	90	113	lm
$\Theta^{1/2}$	50% Power Angle	-	120	-	deg

constellation. We set the CP length three times of the root mean square (RMS) delay spread of the worst impulse response and assume a data with minimum rate of 25 Mbps[2]. The sum of ambient light shot noise and receiver thermal noise is modeled as real baseband additive white Gaussian noise (AWGN) with zero mean and power of -10 dBm. Furthermore, to take LED nonlinearity into account, OPTEK, OVSPxBCR4 1-watt white LED is considered in simulations whose optical and electrical characteristics are given in Table 8.2. A polynomial order of five is used to realistically model measured transfer function. Fig. 8.10 demonstrates the nonlinear transfer characteristics of the LED from the data sheet and using the polynomial function. The four OPTEK LEDs are biased at 3.2 V.

8.4.1 PERFORMANCE COMPARISON OF SINGLE- AND MULTICARRIER MODULATION SCHEMES

In this sub-section, the positioning performance of the OFDM system is compared with the performance of those using single carrier modulation schemes, i.e., on-off keying (OOK). We assume that the average electrical power of the transmitted signal before modulating each LED is $P_{t_e,k} = 5$ dBm. For OOK modulation, the positioning algorithm discussed in Chapter 4 is used[3].

Fig. 8.11 demonstrates the positioning error distribution over the room for an indoor OFDM VLC system with 4-QAM modulation and the FFT size of 512. As it can be seen, the positioning errors are very small for most locations inside the room but become larger when the receiver approaches the corners and edges due to the severity of the multipath reflections.

Fig. 8.12, on the other hand, shows the positioning error distribution over the room for an indoor VLC system employing OOK modulation with the same data rate as that of the OFDM system with 4-QAM modulation (i.e., 25 Mbps). As observed, the

---

[2]The data rate is 25 Mbps for 4-QAM modulation scheme. For higher-order (i.e., $M > 4$), the bit rate can be achieved as $R = 25$ Mbps $\times \log_2(M)$.

[3]The positioning algorithm used for OOK is quite similar to the one used for OFDM. The two algorithms only differ in the estimation method of the signal attenuation.

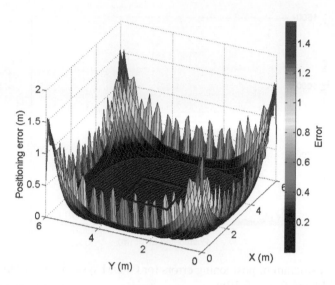

Figure 8.11: Positioning error distribution for OFDM system with 4-QAM modulation, $N = 512$ and $P_{t_e,k} = 5$ dBm.

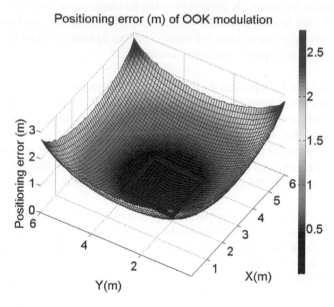

Figure 8.12: Positioning error distribution for OOK modulation with $P_{t_e,k} = 5$ dBm.

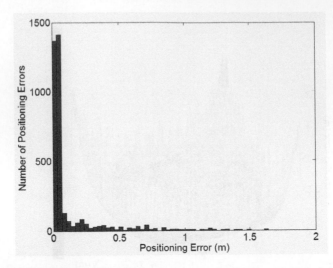

Figure 8.13: Histogram of positioning errors for OFDM system with 4-QAM modulation, $N = 512$ and $P_{t_e,k} = 5$ dBm.

positioning errors are relatively small within the rectangle shown in Fig. 8.12 where the LED bulbs are located right above its corners. However, the positioning error becomes significantly larger when the receiver moves toward the corners and edges as the effect of the multipath reflections increases.

Fig. 8.13 and Fig. 8.14 present the histograms of the positioning errors for OFDM and OOK modulation schemes, respectively. For OFDM modulation, most of the positioning errors are less than 0.1 m and only a few of them are more than 1 m corresponding to the corner area. However, for OOK modulation, the positioning errors are widely spread from zero to around 2.3 m, and only a few of them are less than 0.1 m that correspond to the central area. From Fig. 8.11 to Fig. 8.14, it can be clearly seen that the OFDM system outperforms its OOK counterpart.

Table 8.3 summarizes and compares the positioning errors of OFDM and OOK modulation schemes. As seen, OFDM modulation provides a much better positioning accuracy than OOK modulation for all the locations inside the room. Particularly, the RMS error is 0.08 m for the rectangular area covered perfectly by the four LED bulbs when OFDM modulation is used, while it is 0.43 m for OOK modulation. The total RMS errors are 0.2609 m and 1.01 m for OFDM and OOK modulation schemes as the rectangular area covered by LED bulbs is only 11.1% of the total area. Thus, OFDM modulation decreases the RMS error by 74% compared to OOK modulation. It should be noted that the average positioning accuracy can be increased by optimizing the layout design of the LED bulbs in the future.

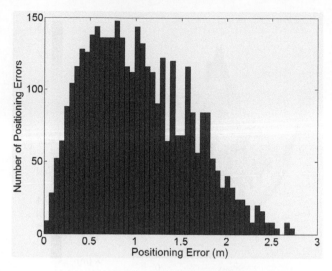

Figure 8.14: Histogram of positioning errors for OOK modulation with $P_{t_e,k} = 5$ dBm.

Table 8.3

Positioning error for single- and multicarrier modulation schemes

Positioning error (m)	OFDM modulation (m)	OOK modulation (m)
Corner (0, 0)	0.578	2.18
Edge (3 m, 0)	0.49	1.53
Center (3 m, 3 m)	2×10^{-6}	10^{-5}
RMS Error of the Rectangular Area	0.08	0.43
RMS Error of the Whole Room	0.2609	1.01

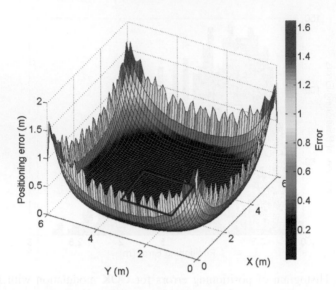

Figure 8.15: Positioning error distribution for OFDM system with 4-QAM modulation, $N = 512$ and $P_{t_e,k} = -10$ dBm.

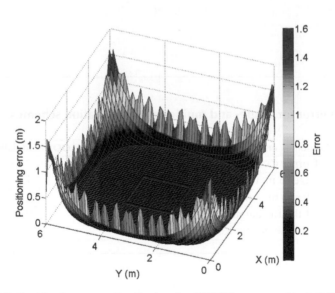

Figure 8.16: Positioning error distribution for OFDM system with 4-QAM modulation, $N = 512$ and $P_{t_e,k} = 20$ dBm.

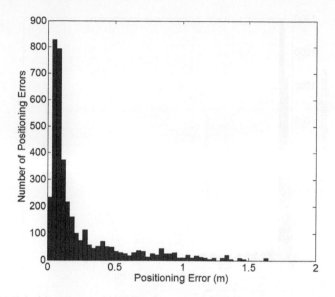

Figure 8.17: Histogram of positioning errors for OFDM system with 4-QAM modulation, $N = 512$ and $P_{t_e,k} = -10$ dBm.

8.4.2 EFFECT OF SIGNAL POWER ON THE POSITIONING ACCURACY

In this sub-section, we investigate the effect of the average electrical power of the transmitted signal on the positioning accuracy of the proposed OFDM VLC system. We consider an OFDM system with 4-QAM modulation and $N = 512$. Fig. 8.15 and Fig. 8.16 present the positioning error distribution for the OFDM system with transmitted signal power values of -10 dBm and 20 dBm, respectively. The total RMS error is calculated as 0.384 m for the OFDM system with $P_{t_e,k} = -10$ dBm and 0.2766 m for $P_{t_e,k} = 20$ dBm.

Fig. 8.17 and Fig. 8.18 further demonstrate the corresponding histograms of the positioning errors for different transmitted signal power values. It is apparent from Fig. 8.15 and Fig. 8.17 that the OFDM positioning system works satisfactorily even at very low transmitted signal power values resulting from dimming and shadowing effects[4]. Furthermore, according to Fig. 8.11, Fig. 8.13 and Fig. 8.15 to Fig. 8.18 and as expected, increasing the average electrical power of the transmitted signal results in a better performance. However, at very high power values, nonlinearity distortion effects dominate the performance and the positioning accuracy decreases. It is the main reason the performance of the VLC system with $P_{t_e,k} = 20$ dBm is slightly worse than that of $P_{t_e,k} = 5$ dBm presented earlier. Therefore, for an OFDM indoor VLC positioning system, there is an optimum power value that depends on the LED characteristics.

---

[4]Note that the total RMS error for OOK modulation with $P_{t_e,k} = -10$ dBm is calculated as 1.32 m.

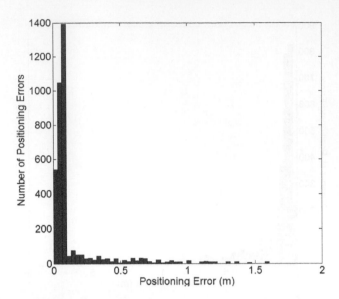

Figure 8.18: Histogram of positioning errors for OFDM system with 4-QAM modulation, $N = 512$ and $P_{t_e,k} = 20$ dBm.

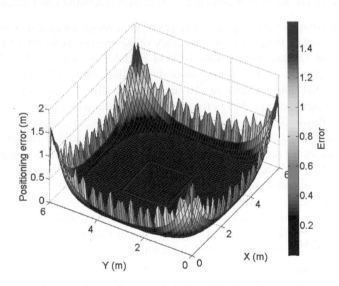

Figure 8.19: Positioning error distribution for OFDM system with 16-QAM modulation, $N = 512$ and $P_{t_e,k} = 5$ dBm.

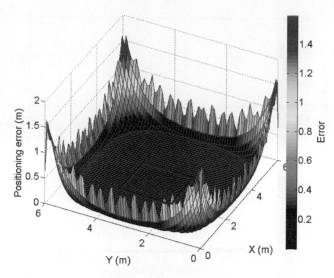

Figure 8.20: Positioning error distribution for OFDM system with 64-QAM modulation, $N = 512$ and $P_{t_e,k} = 5$ dBm.

8.4.3   EFFECT OF MODULATION ORDER ON THE POSITIONING ACCURACY

Here, we analyze the impact of the modulation order on the positioning performance. Figs. 8.19 and 8.20 show the positioning error distribution of the OFDM system with the FFT size of 512 and $P_{t_e,k} = 5$ dBm employing 16- and 64-QAM modulation, respectively. The corresponding histograms of the positioning errors are shown in Figs. 8.21 and 8.22. The total RMS error is obtained as 0.2665 m for 16-QAM and 0.2716 m for 64-QAM. By comparing these RMS error values with the one calculated for 4-QAM modulation, we observe that all three systems yield nearly the same positioning performance. Thus, the constellation size does not have a significant effect on the positioning performance of the proposed OFDM VLC system although the communication performance obviously deteriorates with increasing the constellation size. The numerical results clearly show that the proposed channel DC gain estimation works perfectly for high-order constellations as well.

8.4.4   EFFECT OF NUMBER OF SUBCARRIERS ON THE POSITIONING ACCURACY

Finally, we investigate the effect of number of total subcarriers (i.e., the FFT size) on the positioning accuracy. We consider an OFDM system with 4-QAM and $P_{t_e,k} = 5$ dBm. Figs. 8.23 to 8.28 illustrate the positioning error distribution for different FFT sizes providing sufficiently narrow-banded sub-channels along with their corresponding error histograms. The total RMS errors are respectively calculated as 0.2905 m, 0.271 m and 0.2624 m for $L = 64$, 256 and 1024.

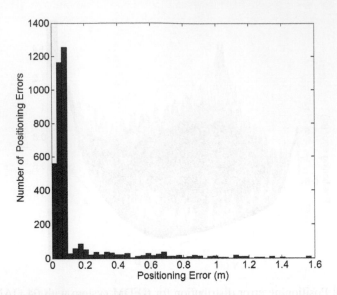

Figure 8.21: Histogram of positioning errors for OFDM system with 16-QAM modulation, $N = 512$ and $P_{t_e,k} = 5$ dBm.

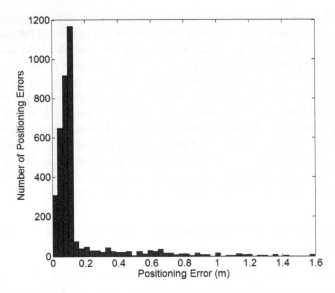

Figure 8.22: Histogram of positioning errors for OFDM system with 64-QAM modulation, $N = 512$ and $P_{t_e,k} = 5$ dBm.

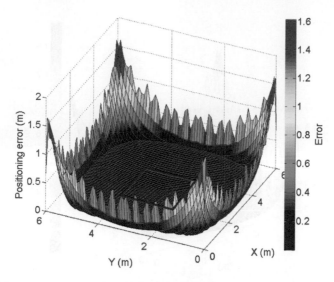

Figure 8.23: Positioning error distribution for OFDM system with 4-QAM modulation, $N = 64$ and $P_{t_e,k} = 5$ dBm.

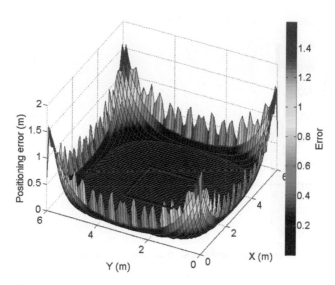

Figure 8.24: Positioning error distribution for OFDM system with 4-QAM modulation, $N = 256$ and $P_{t_e,k} = 5$ dBm.

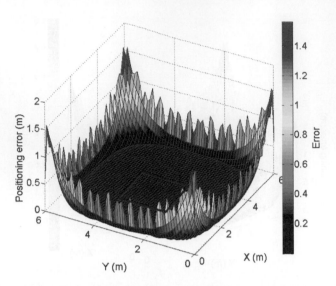

Figure 8.25: Positioning error distribution for OFDM system with 4-QAM modulation, $N = 1024$ and $P_{t_e,k} = 5$ dBm.

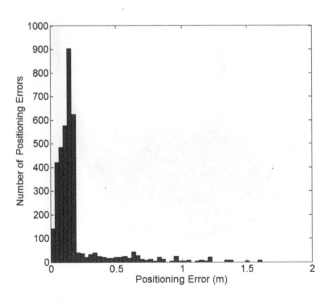

Figure 8.26: Histogram of positioning errors for OFDM system with 4-QAM modulation, $N = 64$ and $P_{t_e,k} = 5$ dBm.

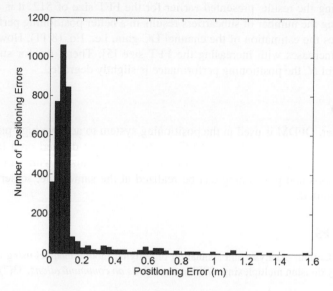

Figure 8.27: Histogram of positioning errors for OFDM system with 4-QAM modulation, $N = 256$ and $P_{t_e,k} = 5$ dBm.

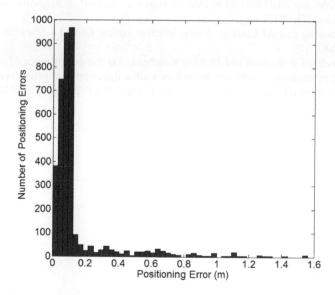

Figure 8.28: Histogram of positioning errors for OFDM system with 4-QAM modulation, $N = 1024$ and $P_{t_e,k} = 5$ dBm.

Considering the results presented earlier for the FFT size of 512, it is observed that increasing the number of subcarriers results in a better positioning performance as it improves the estimation of the channel DC gain, i.e., Eq. (8.11). However, the PAPR also increases with increasing the FFT size [5]. Therefore, for sufficiently large values of N, the positioning performance is slightly degraded.

SUMMARY

In this chapter, OFDM is used in the positioning system to achieve high positioning accuracy. The prior arts on indoor VLC positioning were defined on a low-speed modulation, while high data rate transmission can be realized with OFDM. Both communications and positioning can be realized at the same time, so service data can be transmitted.

REFERENCES

1. Leonard Cimini. Analysis and simulation of a digital mobile channel using orthogonal frequency division multiplexing. *IEEE transactions on communications*, 33(7):665–675, 1985.
2. Sian C Jeffrey Lee, Sebastian Randel, Florian Breyer, and Antonius MJ Koonen. PAM-DMT for intensity-modulated and direct-detection optical communication systems. *IEEE Photonics Technology Letters*, 21(23):1749–1751, 2009.
3. Sarangi Devasmitha Dissanayake and Jean Armstrong. Comparison of ACO-OFDM, DCO-OFDM and ADO-OFDM in IM/DD systems. *Journal of lightwave technology*, 31(7):1063–1072, 2013.
4. Jean Armstrong and AJ Lowery. Power efficient optical OFDM. *Electronics letters*, 42(6):1, 2006.
5. Mohammadreza A Kashani and Mohsen Kavehrad. On the performance of single-and multi-carrier modulation schemes for indoor visible light communication systems. In *2014 IEEE Global Communications Conference*, pages 2084–2089. IEEE, 2014.

9 Sensor Fusion

Sensor fusion is combining of sensory data or data derived from disparate sources such that the resulting information has less uncertainty than would be possible when these sources were used individually. The term *uncertainty reduction* in this case can mean more accurate, more complete, or more dependable, or refer to the result of an emerging view, such as stereoscopic vision.

The data sources for a fusion process are not specified to originate from identical sensors. One can distinguish direct fusion, indirect fusion and fusion of the outputs of the former two. Direct fusion is the fusion of sensor data from a set of heterogeneous or homogeneous sensors, soft sensors, and history values of sensor data, while indirect fusion uses information sources like a priori knowledge about the environment and human input.

An inertial navigation system (INS) is a navigation assistant that uses an inertial measurement unit (IMU), typically composed of accelerometers and gyroscopes, to continuously calculate the position, orientation, velocity and trajectory of a moving object. The INS available in smart phones and wearable devices frequently suffers from big positional errors in trajectories of motions and long-range displacements due to less-accurate IMU outputs—especially gyroscopes. Innovative technologies could improve the performance of INS with the advancement of IMU units by new physical methods, unique materials, and innovative fabrication techniques and/or with (i) the addition of different sensors, (ii) the sensor data fusion of external sensing and/or detection technologies, or (iii) smart computational algorithms for specific applications. We believes INS with high accuracy can create new values for consumer electronics.

Solid-state lighting (SSL) has the potential to significantly reduce lighting energy use and slash greenhouse-gas emissions. The Department of Energy (DoE) [1] estimates that switching to light emitting diode (LED) lighting over the next 20 years could save $250 billion in energy costs over that period, reduce electricity consumption for lighting by nearly one-half, and avoid 1,800 million metric tons of carbon emissions. The energy-saving promise of SSL technology has particular market relevance given the ongoing transition to higher-efficiency bulbs, as mandated by the Energy Independence and Security Act of 2007. The U.S. The DoE also states that LEDs had 4 percent of the U.S. lighting market in 2013, but it predicts this figure will rise to 74 percent of all lights by 2030 [1]. Though SSL is still at an early stage of development, it is evolving rapidly, with new generations of devices introduced every few months. Many of these products can save energy and provide high quality lighting in a growing number of applications and their overall quality is improving steadily. Cost is another major consideration in evaluating LED lighting. Today, quality LED products cost more than conventional lighting products at the outset, yet lifecycle savings can outweigh the initial cost premium in applications that take advantage of LED's unique features – including directionality, controllability, long lifetimes, and, of course, lower power consumption. With the growth of wireless communication networks, the need for energy-efficient networks is more urgent and pressing than before. Statistics published by the Bell Labs indicate the rapidly accelerating dominance of wireless access power consumption by 2020. It can grow by a factor of 100 in the next 10 years. This means the wireless access power per user will approach 100 Watts, unless the network efficiency in terms of "Total traffic per user/Total power per user" can be increased through power saving methods. Efforts are under way to place information and communications technology (ICT) networks on an innovative path to ensure the sustainability of network growth in future decades, in such a way that ICT can lead in enabling reduction of global energy consumption and carbon footprint.

The introduction of networked lights is happening because of another trend. Manufacturers have been replacing incandescent and fluorescent lights with ultra-efficient LEDs that have an Internet protocol (IP) address. Because LEDs are solid-state devices that emit light from a semiconductor chip, they already sit on a circuit board, so can readily share space with sensors, wireless chips, and a small computer, allowing light fixtures to become networked sensor hubs, thus "Internet of Light." LED trends can precipitate a revolution in efficiency and brightness and the fact that these can work with many of the pre-existing infrastructures is a very convenient ubiquitous feature. Another outstanding advantage of LED is its rather short-time response; therefore, it can be used for high-speed data transmissions, thus enabling visible light communication (VLC), sensing, navigation, positioning, etc.

9.1 METHODS AND GOALS

A very realistic problem with any positioning system, be it guided by radio frequency (RF) or light, can be presented in the form of a simple question. What will happen if no RF/Light guiding signal is available in a certain indoor/outdoor environment, or alternatively, the guiding signals are turned off? Moreover, guiding signal, carrying position information, itself can be blocked by different objects or due to weather conditions. Any of these cases, if not properly addressed, can bring service outage in the positioning system.

In this chapter, we focus on sensor fusion technology that utilizes INS to overcome the temporary service outage due to a blockage. A realization by Kalman filter is detailed and preliminary simulation results are presented. In this context, we choose positioning by light, as global positioning system (GPS) signals do not penetrate indoors well. However, INS designs will be general and can be applied elsewhere, as well. In the future, we will have a lighting source integrated in it, RF as well as infrared signal generators and sensors. More precision in tracking is the point of differentiation from current state-of-the-art. Another objective of this chapter is to determine practical means of improving performance of visible light communication links and quantify effectiveness of proposed efficient design concepts. Optimization principles take into account not only the communications and sensing requirements, but also the efficient energy utilization aspects of lighting LEDs and the quality of light produced. Challenges existing in joint design and optimization of communications and illumination methods must be addressed and properly overcome. The integration capabilities of communications and sensing would significantly enhance the utility of the lighting system and would allow costs to be spread over a wider level of utility. Ubiquitous communications with lighting will help large-scale energy savings and stimulate LED lighting and communication industry as well as substantial economic growth in this new area. There is no doubt that this "green" technology has the potential of answering the problems and technical challenges mentioned earlier. At the same time, there is significant original research needed in order to bring this technology to the level of its full potential.

9.2 INERTIAL NAVIGATION SYSTEM

Inertial Navigation System (INS) is a self-contained navigation technique that utilizes measurements from one or all the following types of sensors: motion sensors (accelerometers) and rotation sensors (gyroscopes) and compasses. INS estimates the relative position to a known initialization, orientation and velocity (travelling direction and speed) of an object.

Normally, three accelerometers and three gyroscopes are combined to make an IMU, which is sufficient to navigate an object through 3-D space, given initial values of space position and velocity. Nearly every IMU can be categorized into one of the following two kinds: gimbaled or stable platform systems and strap-down systems [2].

9.2.1 STABLE PLATFORM SYSTEMS

In stable platform systems, inertial sensors are mounted on a platform that is held by gimbals. A gimbal is a rigid frame with rotation bearings which isolates the inside platform from external rotations while gives the platform freedom in all three axes, as shown in Fig. 9.1 [2]. Gyroscopes mounted on the inside platform detect any platform rotations, which are fed back to torque motors. Torque motors rotate the gimbals according to feedback signals in order to cancel platform rotations, keeping the platform stable. As we can see, gimbals in a typical IMU are mounted one inside another and at least three gimbals are required to isolate the sensors from rotations in three axes.

To track the orientation of the device, angles between gimbals can be obtained using angle pick-offs. The position of the device is computed utilizing signals from accelerometers mounted on the platform, which are double integrated to get travel distance information. Note that it is a must to subtract gravity from the vertical channel's signal before the integration. Algorithm for stable platform inertial navigation is shown in Fig. 9.2 [2].

9.2.2 STRAP-DOWN SYSTEMS

In strap-down systems, inertial sensors are mounted directly on devices, vehicles, airplanes etc. as shown in Fig. 9.3 [3]. Therefore, the measurements are with respect

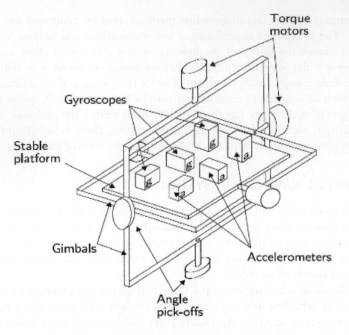

Figure 9.1: A stable platform unit [1].

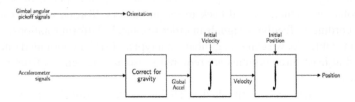

Figure 9.2: Stable platform inertial navigation algorithm [1].

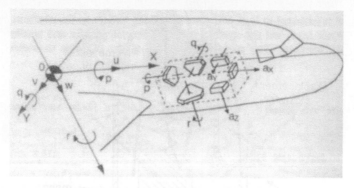

Figure 9.3: Example of strap-down inertial navigation system on airplane [2].

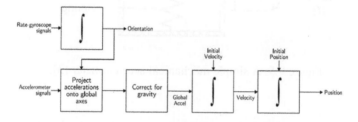

Figure 9.4: Strap-down inertial navigation algorithm [1].

to the body frame rather than global frame. To obtain orientation signals, gyroscopes are integrated. Acceleration measurements from accelerometers have to be translated from body coordinates into global coordinates, which is done by direction cosine matrix (DCM) determined by the integration of gyroscope signals. Position information is then obtained by integrating global acceleration signals as in the stable platform algorithm. This procedure is shown in Fig. 9.4 [2].

Compared to stable platform systems, strap-down solutions provide reduced mechanical complexity that means smaller size and better reliability can be reached. These advantages are tradeoffs at the price of higher computational complexity. However, since the cost of computation has dropped dramatically in the last two decades, strap-down systems have become dominant.

9.3 INERTIAL SENSORS

It should be noticed that due to weight and size issues, inertial sensors have not been used in many scenarios. However, recent improvements of microelectromechanical systems (MEMS) inertial sensors have brought opportunities to vast applications of INS, especially in consumer electronics.

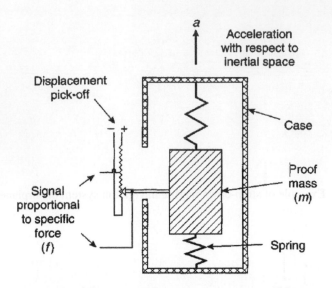

Figure 9.5: A simple mechanical accelerometer [3].

As mentioned earlier, typically an IMU is composed of three accelerometers and three gyroscopes. In this sub-section, we will review MEMS type of these two kinds of inertial sensors and brief their measurement error characteristics.

9.3.1 MEMS ACCELEROMETER

There are two major classes of MEMS accelerometers. The first is called mechanical accelerometers manufactured by MEMS techniques that measure the translational motion of a supported mass as in traditional accelerometer as shown in Fig. 9.5 [4]. The second class refers to devices that make measurement of the frequency change of a vibrating element as tension changes. For a complete survey, the reader should refer to [4].

We consider two types of errors described in [2]: constant bias and thermo-mechanical white noise.

The most important error source of an accelerometer is the bias. The bias of an accelerometer is the deviation of its output signal from the object's true acceleration, in unit of m/s^2. After double integration, a constant bias error of ε is translated into a quadratic error in position that grows with time. The accumulated error in position is:

$$s(t) = \varepsilon \cdot \frac{t^2}{2} \qquad (9.1)$$

where t is the time of the integration.

It is possible to estimate the bias by observing the average of the accelerometer's output given no actual acceleration over the long term. In practice, this calibration is usually finished by mounting the device on a turntable whose orientation can be adjusted accurately.

The output of MEMS accelerometers is also perturbed by thermo-mechanical white noise whose frequency is much higher than the sampling rate of accelerometers. In [2] the authors showed this white noise produces a velocity random walk. Analysis shows that this second order random walk in position has zero mean and a standard deviation given by:

$$\sigma_s(t) \approx \sigma \cdot t^{3/2} \cdot \sqrt{\frac{\delta t}{3}} \tag{9.2}$$

where σ is the standard deviation of the white noise on the accelerometer's output and δt is the time sampling period. The notion of random walk refers to a process composed of successive steps, the size of which, as well as the direction, is unknown.

9.3.2 MEMS GYROSCOPE

Compared to mechanical and optical counterparts, MEMS gyroscopes built using silicon micro-machining techniques have simpler structure that consists of as few as three parts therefore ensuring the reliability as well as reducing the manufacturing cost. Same as for accelerometers, we also show two types of errors for MEMS gyroscopes: constant bias and thermo-mechanical white noise.

The bias of a gyroscope is the average output given no rotation over a long time, in unit of $\circ/h$. After double integration, a constant bias error of ε is translated into a linear error in angle that grows with time. The accumulated error in angle is expressed by:

$$\theta(t) = \varepsilon \cdot t \tag{9.3}$$

where t is the time of the integration.

The output of MEMS gyroscopes is also perturbed by white noise. In [5], authors showed that this white noise creates a zero-mean random walk in the output signal, whose standard deviation is expressed by:

$$\sigma_\theta(t) = \sigma \cdot \sqrt{\delta t \cdot t} \tag{9.4}$$

We can see that this deviation is proportional to the square root of integration time.

9.4 REALIZATION BY KALMAN FILTER

As we can see from last sub-section, one problem with INS is that the positioning errors accumulate over time with increasing variance, generating very large errors if not properly corrected. Fig. 9.6 shows experimental results of INS performance by placing an INS system stationary on a trolley to estimate and compensate the gravity component. The green line represents the actual path and blue dots indicate

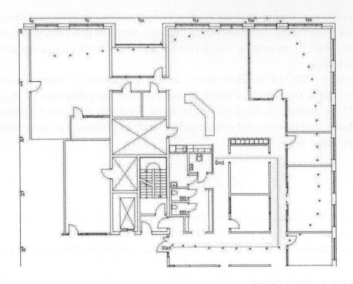

Figure 9.6: Position estimation by using INS only [6].

position estimates given by INS system only. As we may see, the error in estimating acceleration results in an invertible higher speed estimate, which leads to a huge drift from real path after several estimates.

We propose a realization method of sensor fusion combining the output of proposed light positioning system with a 3-axis accelerometer, by incorporating two Kalman filters. The role of the first Kalman filter is to keep an inner state transition equation based on previous state estimates, predict the next state based on it while take estimates from inertial navigation system to update the prediction. The updated state estimate from the first Kalman filter is then passed to the second Kalman filter which further updates it with estimates from our light positioning system (LPS) and passes it to the output. This output is also fed back to the first Kalman filter to update the priori position for INS. To prevent outlier generated in either INS or LPS, we detect possible malfunctioning of both positioning systems by finding abnormal values which lie far from normal output range, i.e., values should be treated as an output. One instance of this is the blockage of line-of-sight (LoS) path for our light positioning system as mentioned previously. We also put a decision maker after Kalman filters comparing the input of the second Kalman filter (LPS estimates) with the first one's output to eliminate the effect of possible outliers in LPS estimates. We set a threshold offset, which is the tolerance of deviation between LPS estimates and the first Kalman filter's output. If the threshold offset is met, the filter will make the first Kalman filter's output final. In the meantime, we maintain a counter for this threshold and if the counter's value is larger than a preset value we will reverse the final output to that of the second Kalman filter.

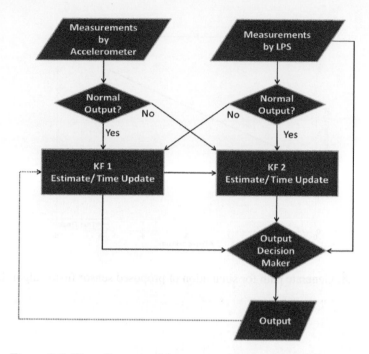

Figure 9.7: Flow diagram of the proposed sensor fusion algorithm.

The whole process may sound complicated but can be shown in a clear manner in Fig. 9.7.

9.5 SIMULATION AND RESULTS

We test a proposed sensor fusion algorithm for a 2-D navigation. The error characteristics of accelerometer in our simulation are generated based on data provided in [2]. The biases and white noises on output of accelerometers downgrade INS performance in the manner discussed previously. Values of these two factors of accelerometers on X and Y axes are provided in Table 9.1.

Table 9.1

Parameters used to characterize accelerometers

	Bias Instability (ε)	White Noise (σ)
X-Accelerometer (m/s^2)	1.2×10^{-6}	0.011
Y-Accelerometer (m/s^2)	2.7×10^{-6}	0.011

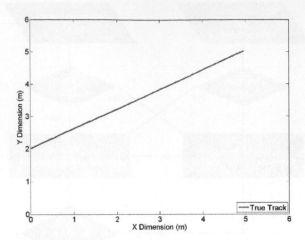

Figure 9.8: Generate path for simulation of proposed sensor fusion algorithm.

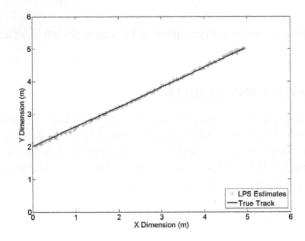

Figure 9.9: Generate path and LPS estimates for simulation of proposed sensor fusion algorithm.

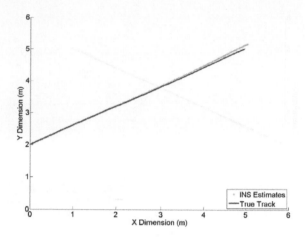

Figure 9.10: Generate path and INS estimates for simulation of proposed sensor fusion algorithm.

Table 9.2

Positioning accuracy of solutions (in RMS value)

	Scenario 1	Scenario 2
INS only	5.96 cm	5.96 cm
LPS only	4.85 cm	N/A
Sensor Fusion	4.02 cm	4.56 cm

We consider a path of the receiver as shown in Fig. 9.8 and generate estimates using both INS and proposed LPS. The INS system is assumed to have a very accurate initialization. Results are shown in Fig. 9.9 through Fig. 9.11 if LPS is able to successfully make estimates on most of the receiver's positions, which we define as Scenario 1. In cases that a multi-access protocol fails, LPS estimates are made equal to the last successful estimate.

As we can see, INS provides us with cumulative positioning error over time, which is shown as an increasing deviation between INS estimates and the real path, which will become even larger as time goes on. Thus though a good initialization may lead to excellent performance, INS cannot yield satisfactory performance by itself. The use of sensor fusion helps to improve positioning accuracy compared to relying on LPS estimates only, which is shown in Table 9.2.

In Scenario 2, we take outlier and light blockage into consideration to show proposed sensor fusion helps to mitigate the effects of both and improve positioning

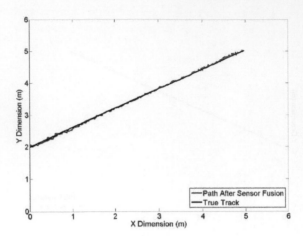

Figure 9.11: Generate path and sensor fusion estimates for simulation of proposed sensor fusion algorithm.

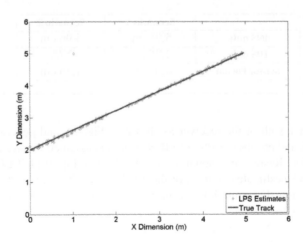

Figure 9.12: LPS estimates and the true track in the presence of outlier and blockage.

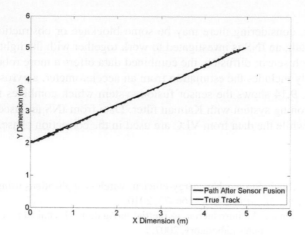

Figure 9.13: Sensor fusion estimates and the true track in the presence of outlier and blockage.

accuracy. In the case that multi-access protocol fails, LPS generates a flag indicating an abnormal output. Results are shown in Figs. 9.12 and 9.13, true track of the receiver and INS-only estimates stay the same as shown in Figs. 9.8 and 9.10, respectively. Numerical results are shown in Table 9.2.

Indeed, we can see that sensor fusion technology offers help in further improving positioning accuracy, even if LPS system yields satisfyingly proper estimates. Moreover, in the presence of outlier and/or light blockage inside real LPS, sensor fusion provides stable and reliable estimates of the receiver's location, taking advantage of both INS and our proposed LPS.

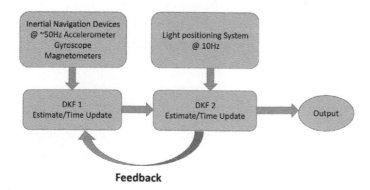

Figure 9.14: Sensor fusion with centralized Kalman filter.

SUMMARY

In this chapter, considering there may be some blockage or obstruction along the transmission path, an INS is investigated to work together with the light positioning system. Through sensor diffusion, the combined data offers a more robust positioning. INS usually includes the estimation from an accelerometer, a gyroscope and an e-campus. Fig. 9.14 shows the sensor fusion system which combines the INS and the light positioning system with Kalman filter. Data from INS are used in the time update phase, while the data from VLC are used in the correction phase.

REFERENCES

1. Mohsen Kavehrad. Sustainable energy-efficient wireless applications using light. *IEEE Communications Magazine*, 48(12):66–73, 2010.
2. Oliver J Woodman. An introduction to inertial navigation. Technical report, University of Cambridge, Computer Laboratory, 2007.
3. Basic principles of inertial navigation, seminar on inertial navigation systems, Tampere University of Technology. http://atlas.physics.arizona.edu/~kjohns/downloads/inertial/InertialNavigationSystems.pdf.
4. David Titterton, John L Weston, and John Weston. *Strapdown inertial navigation technology*, volume 17. IET, 2004.
5. Christopher M Bishop. *Pattern recognition and machine learning*. Springer, 2006.
6. Ubejd Shala and Angel Rodriguez. Indoor positioning using sensor-fusion in android devices, p. 58, 2011.

Glossary

ACO-OFDM asymmetrically clipped optical OFDM. 109, 111, 112, 114, 151
AI artificial intelligence. 1, 151
AOA angle of arrival. 20, 21, 27, 61, 66, 70, 71, 73, 75, 78, 151
AWGN additive white Gaussian noise. 119

BFSA basic framed slotted ALOHA. xiii, 31, 33, 42, 45, 56, 151
BLE Bluetooth low energy. 23, 151
BOF beginning of flag. 33, 151
BPPM binary pulse position modulation. 13, 151

CDF cumulative distribution function. xvi, 36, 42, 57–59, 64, 93, 99, 100, 104, 151
CDMMC combined deterministic and modified Monte Carlo. 81, 85, 86, 104, 114, 151
CFL compact fluorescent light. xii
CP cyclic prefix. 110–112, 114, 117, 151
CPC compound parabolic concentrator. 12, 36, 70, 151
CRC cyclic redundancy check. 33, 151
CT compensation time. 13, 151

DC direct current. 12, 28, 72–75, 83, 85, 109, 111, 114, 116, 126, 130, 151
DCM direction cosine matrix. 137, 151
DCO-OFDM DC-clipped optical OFDM. 109, 111, 151
DD direct detection. 12, 13, 109, 151
DoE Department of Energy. 133
DPIM digital pulse interval modulation. 14, 151

EM electro-magnetic. 26
EOF ending of flag. 33, 151

FCS frame correction sequence. 33, 151
FDM frequency-division multiplexing. 107, 151
FET field-effect transistor. 15, 151
FFT fast Fourier transform. 107, 110, 112, 119, 126, 130, 151
FOV field-of-view. 12, 13, 21, 28, 36, 61, 72, 81, 89, 114, 117, 151

GI guard interval. 107, 151
GM-SPPF Gaussian mixture sigma-point particle filter. xiii, 3, 28, 45, 54, 56, 59, 151
GMM Gaussian mixture model. 51, 52, 54, 151
GPS global positioning system. xi, 1, 3, 18, 22, 23, 28, 134, 151

ICT information and communications technology. 134, 151
ID identification. 23, 27, 31, 33, 35, 42, 114, 117, 156
IFFT inverse fast Fourier transform. 107, 110–112, 151
IM intensity modulation. 12, 13, 109, 151
IMU inertial measurement unit. xiii, 25, 26, 133, 135, 138, 151
INS inertial navigation system. xiii, 25, 26, 133–135, 137, 139–141, 143, 145, 146, 151
IP Internet protocol. xii, 134, 151
IR infrared. 5, 151
ISI intersymbol-interference. 107, 109, 151
ISM industrial, scientific and medical. 23, 151

LBS location based service. 1, 2, 151
LED light emitting diode. xii, 5, 6, 9, 19, 22, 26–28, 31, 33, 34, 36, 48, 67, 69–72, 75, 85, 86, 91, 92, 100, 102, 104, 109, 112, 114, 116, 117, 119, 120, 122, 133–135, 151
LLSE linear least square estimation. 35, 67, 95, 99, 151
LoS line-of-sight. xii, xiii, 3, 17, 23, 28, 56, 67, 69–72, 75, 77, 78, 81, 83, 85, 86, 89, 91–93, 114, 140, 152, 153
LPS light positioning system. 140, 143, 145, 152
LQE linear quadratic estimation. 45, 152
LTI linear time-invariant. 81, 152

MEMS microelectromechanical systems. 137–139, 152
MMC modified Monte Carlo methods. 81, 85, 86, 152

OFDM orthogonal frequency-division multiplexing. xiii, 3, 4, 107, 109, 111, 112, 114, 116, 117, 119, 120, 122, 126, 130, 151, 152
OOK on-off keying. 13, 31, 119, 120, 122, 151, 152

P/S parallel to serial. 111
PAM-DMT pulse-amplitude-modulated discrete multitoned. 109, 152
PAPR peak-to-average-power ratio. 107, 109, 111, 130, 152
PD photodiode. 8, 9, 12, 13, 15, 27, 28, 34, 42, 48, 61, 71, 81, 83, 85, 86, 89, 114, 117, 152
PDA personal digital assistant. 23, 152
PDF probability density function. 51, 52, 85, 86
PPM pulse position modulation. 13, 14, 152
PSD power spectral density. 24, 152
PSK phase-shift keying. 107, 152

QAM quadrature amplitude modulation. 107, 112, 114, 152

RF radio frequency. xi, 3, 15, 20, 21, 24, 26, 27, 70, 134
RFID radio frequency identification. 3, 20, 22–24, 152

RMS root mean square. 3, 18, 36, 41, 42, 59, 64, 93, 99, 100, 102, 117, 120, 122, 126, 152, 157
RSS received signal strength. 20, 21, 25–27, 42, 66, 70, 71, 73, 75, 78, 91, 152
RSSI received signal strength indication. 23

SNR signal-to-noise ratio. 9, 152
SPA sigma-point approach. 52
SSL solid-state lighting. 133, 152

TDMA time division multiple access. 31, 33, 42, 152
TDOA time-difference-of-arrival. 19, 21, 24, 25, 27, 152
TOA time of arrival. 18–21, 24, 27, 152

UV ultraviolet. 5, 152
UWB ultra-wideband. 19, 20, 24, 25, 28, 152

VL visible light. xii
VLC visible light communication. xii, xiii, xv, 3–6, 8, 9, 12, 15, 17, 19, 20, 26–28, 31, 33, 34, 45, 64, 66, 75, 107, 109, 117, 119, 120, 122, 126, 130, 134, 146, 152
VPPM variable pulse position modulation. 13, 14, 152

WLAN wireless local area network. 3, 25, 26, 28, 107, 152

Index

asymmetrically clipped optical
OFDM (ACO-OFDM),
106, 109, 111
artificial intelligence (AI), 1
angle of arrival (AOA), 20, 21, 27,
61, 65, 69, 70, 74, 77
basic framed slotted ALOHA
(BFSA), 31, 33, 42, 45, 56
Bluetooth low energy (BLE), 23
beginning of flag (BOF), 33
binary pulse position modulation
(BPPM), 13
cumulative distribution function
(CDF), 36, 42, 57–59, 64,
91, 97, 98, 102
combined deterministic and modified
Monte Carlo (CDMMC),
79, 83, 84, 102, 111
compound parabolic concentrator
(CPC), 12, 36, 69
cyclic prefix (CP), 108, 110, 111, 116
cyclic redundancy check (CRC), 33
compensation time (CT), 13
direction cosine matrix (DCM), 133
DC-clipped optical OFDM
(DCO-OFDM), 106, 108,
109
direct current (DC) biasing, 111
direct detection (DD), 12, 13, 106
Department of Energy (DoE), 129
digital pulse interval modulation
(DPIM), 14
ending of flag (EOF), 33
frame correction sequence (FCS), 33
frequency-division multiplexing
(FDM), 105
field-effect transistor (FET), 15
fast Fourier transform (FFT), 105,
108, 110, 116, 123, 128

field-of-view (FOV), 12, 13, 21, 36,
61, 71, 79, 87, 111, 114
guard interval (GI), 105
Gaussian mixture model (GMM), 51,
52, 54
Gaussian mixture sigma-point
particle filter (GM-SPPF),
45, 51, 54, 56, 59
global positioning system (GPS), 1,
3, 18, 22, 23
information and communications
technology (ICT), 130
inverse fast Fourier transform (IFFT),
105, 107, 108, 110
inertial measurement unit (IMU),
129, 131, 134
intensity modulation (IM), 12, 13, 106
inertial navigation system (INS), xiii,
25, 26, 129–131, 133,
135–137, 139, 141, 142
Internet protocol (IP) address, 130
infrared (IR), 5
intersymbol-interference (ISI), 105,
106
industrial, scientific and medical
(ISM), 23
location based service (LBS), 1, 2
light emitting diode (LED), 5–7, 9,
19, 22, 26, 27, 31, 33, 34,
36, 48, 67, 68, 70, 71, 74,
84, 89–91, 97–102, 106,
109–111, 114–116, 118,
121, 129–131
LED nonlinearity, 116
linear least square estimation
(LLSE), 35, 67, 93, 97
line-of-sight (LoS), 74
light positioning system (LPS), 136,
139, 141

linear quadratic estimation (LQE), 45
linear time-invariant (LTI), 79
microelectromechanical systems (MEMS), 133–135
modified Monte Carlo methods (MMC), 79, 83, 84
orthogonal frequency-division multiplexing (OFDM), xiii, 3, 105, 106, 108, 110, 111, 114–116, 118, 121, 123, 128
on-off keying (OOK), 13, 31, 116, 118
pulse-amplitude-modulated discrete multitoned (PAM-DMT), 106
peak-to-average-power ratio (PAPR), 105, 106, 109, 128
personal digital assistant (PDA), 23
photodiode (PD), 8, 9, 12, 13, 15, 27, 34, 42, 48, 61, 70, 79, 81, 83, 84, 87, 111, 114
pulse position modulation (PPM), 13, 14
power spectral density (PSD), 24
phase-shift keying (PSK), 105
quadrature amplitude modulation (QAM), 105, 109, 111
radio frequency identification (RFID), 3, 21–24
root mean square (RMS) delay spread, 116
RMS error, 3, 18, 36, 41, 42, 59, 64, 91, 97, 98, 100, 118, 121, 123
received signal strength (RSS), 18, 20, 21, 26, 27, 42, 65, 67, 69, 70, 74, 77, 89
signal-to-noise ratio (SNR), 9
solid-state lighting (SSL), 129
time division multiple access (TDMA), 31, 33, 42
time-difference-of-arrival (TDOA), 19, 21, 24, 25, 27

time of arrival (TOA), 18–21, 24, 27
ultraviolet (UV), 5
ultra-wideband (UWB), 19, 20, 24, 25, 28
visible light communication (VLC), 3–6, 8, 9, 12, 15, 17, 19, 20, 26–28, 31, 33, 34, 45, 65, 74, 105, 106, 110, 115, 116, 121, 123, 128, 130, 142
variable pulse position modulation (VPPM), 13, 14
wireless local area network (WLAN), 3, 26, 105
2-D positioning system, 31
3-D positioning system, 31, 41

digital audio broadcasting, 105

abnormal output, 141
absolute temperature, 15
accelerometer, xiii, 75, 129, 131, 133–137, 142
accumulated error, 134, 135
aircraft, 26, 45
airports, 1
ambient light, 14, 116
ambient radiation, 9
amplification circuit, 8
angulation, 17, 20, 27, 69
APD120A2, 9
approximation error, 98
array of antennas, 20
array of e-compasses, 24
assisted-GPS, 3, 22, 23
asynchronous protocol, 31
atmospheric turbulence, 5

background current, 14, 15
bandpass filtering, 9, 12
bandwidth, 5, 6, 8, 9, 14, 15, 24, 25, 33, 42, 105
Barry's algorithm, 82
baseband, 12, 116
beam profiles, 22

Bell Labs, 130
blending factor, 46
blockage, 56, 130, 136, 139–142
blockage of LoS link, 56
Bluetooth, 3, 22, 23, 26
body frame, 133
Boltzmann's constant, 15
brightness, 8, 130

capacitance of photo detector, 15
carbon emissions, 129
carbon footprint, 130
centimeters, 19
channel access methods, 31, 42
chip, 5, 130
circuit board, 130
circular lateration, 18
clipping noise, 106, 108, 109
computation time, 82, 83
computational cost, 52, 70, 82, 83
concentrator gain, 13
confidence interval error, 3, 36, 57,
        59, 64, 97, 100, 102
constant bias, 134, 135
constellation size, 123
consumer electronics, 129, 133
control of vehicles, 45
controllability, 129
convergence, 52, 61, 62, 100
cost function, 62
covariance, 46, 47, 52
Cramer-Rao bound, 19, 27
cyclic prefix, 105

damping factor, 62
data structure, 33
dense grid of reference points,
        22
design robustness, 3
detection area, 8
detection technologies, 129
deterministic approach, 81
diagonal scaling matrix, 62
dielectric CPC, 11
digital video broadcasting, 105

dimming, 13, 14, 121
direct fusion, 129
directional antenna, 20
discrete-data linear filtering, 45
disparate sources, 129
distance estimation, 18, 20, 27, 90,
        98
Doppler shift, 105
driver circuit, 31, 111
dynamic range, 106, 111

econometrics, 45
economic growth, 131
efficient energy utilization, 130
electricity consumption, 129
electromagnetic spectrum, 5
electron charge, 9
emitting elementary area, 8
energy-saving, 129
Ephemeris, 18
equalization, 105
equivalent noise bandwidth, 14
Euclidean norm, 36
exhibition halls, 1
external rotations, 131

fabrication, xiii, 5, 129
fading, 105
Family Locator, 1
feedback signals, 131
feedback-resistor noise, 15
filtering techniques, 3, 28, 45, 57, 59,
        61, 100
fingerprint, 21, 24, 26
flow diagram, 38, 40, 48, 49, 51, 62,
        63, 81–85, 109, 137
fluorescent, 26, 27, 130
fly-eye receiver, 21
forward voltage, 116
Foursquare, 1
frame period, 31
frame structure, 31–33
frequency band, 5
frequency domain, 24, 80
frequency synchronization, 105

frequency time, 79
front-facing cameras, 21

gate leakage current, 15
Gaussian distribution, 14, 46, 49
Gaussian random variable, 39
Geo-Magnetism, 24
geometric properties, 17
gimbaled, 131
global energy consumption, 130
global frame, 133
gradient, 62
gravity component, 135
greenhouse-gas emissions, 129
Groupon, 1
guidance, 2, 45
guiding signal, 130
gyroscope, xiii, 70, 75, 129, 131,
        133–135, 142

health care, 1
hemispherical concentrator, 12
Hermitian symmetry, 106, 107
Hessian matrix, 62
heterogeneous, 129
high quality lighting, 129
homogeneous, 129
horizontal distance, 34, 39, 40, 114
hyperbola, 19
hyperbolic lateration, 19

imaging receiver, 21
importance density, 50
impulse modulation, 24
impulse response, 79–81, 83, 84,
        87–89, 102, 111–114, 116
incandescent, 130
incident angles, 17, 20
indirect fusion, 129
indoor localization, 1, 3
indoor map, 2
indoor navigation, 2
indoor wireless communications, 79
infrared signal generators, 130
innovative technologies, 129

input vector, 46
intensity, 6, 8, 12, 13, 31, 89, 100,
        109–111
inter-source interference, 22
Internet of Light, 130
inverse Fourier transform, 79
iteration process, 50

Jacobi matrix, 62
joint probability distribution, 45

K-means algorithm, 54
K-nearest neighbors algorithm, 24,
        26
Kalman filter, 28, 45–49, 56, 57, 59,
        130, 135, 136, 141, 142

Lambertian order, 6, 71, 84
LANDMARC, 23
large deviations, 28, 45, 59
lateration technique, 35
Levenberg-Marquardt method, 62
lifecycle savings, 129
light speed, 18
lighting bulbs, 31
lighting energy, 129
lighting infrastructure, 21
linear blending, 46
linear convolution, 108
linear power amplifier, 106
location data of a user, 1
long lifetimes, 129
low power consumption, 6
luminance efficiency, 6
luminous flux, 8, 116

manufacturing, 1
measurement noise covariance, 46
mobile apps, 1
mobile device, 1, 67
modulated signal, 12, 111
monochromatic, 6
Monte Carlo estimation, 49
Monte Carlo localization algorithm,
        24

Monte Carlo ray-tracking, 82
motion sensors, 131
motor commands, 45
multi-access protocol, 139, 141
multipath propagation, 79, 105
multipath reflections, 3, 15, 21, 27,
        79, 87, 89–91, 93, 98, 102,
        105, 106, 111, 114, 116,
        118
museums, 1

nanoseconds, 19
narrow-band, 105, 123
navigation, 1, 2, 18, 25, 45, 129–133,
        136, 137
navigation messages, 18
noise bandwidth factor, 15
noise characteristic, 14, 16
non-Gaussian, 49
nonlinear estimation, 3, 61, 64, 93,
        95–99, 101, 102
nonlinear function, 61
nonlinear process, 52
nonlinearity distortion effect, 121

observation noise, 46
observation parameters, 46
observation vector, 46
open-loop voltage gain, 15
optical concentrator, 8–10
optical filter, 8–11, 13
orientation, xiii, 6, 25, 26, 65, 70–73,
        75, 129, 131, 133, 135
orthogonal, 105, 106
outdoor environmental degradations,
        5
outdoor environments, 1
outlier, 45, 100, 136, 139–141

parabolic hollow CPC, 11
partial GPS receiver, 23
Particle filter, 28, 45, 49–52, 56, 57,
        59
passive tags, 23
path loss, 9, 20

photo current, 8
photons, 6–8
Planck's constant, 9
planer filter, 9
platform freedom, 131
polynomial order, 116
position information, 1
positioning accuracy, 3, 15, 19–21,
        26–28, 41, 79, 89–91, 93,
        98, 100, 102, 118, 121,
        123, 128, 139, 141
positioning algorithms and systems,
        17
positioning error, 18, 22, 28, 36, 37,
        41, 42, 57, 59, 64, 72,
        75–77, 89–102, 116–127,
        135, 139
positioning performance, 3, 31, 36,
        45, 56, 61, 64, 89–91, 97,
        100, 102, 116, 123, 128
positive scalar, 62
posteriori, 46
pre-existing infrastructures, 130
predetermined threshold, 50, 52
priori, 46, 47, 129, 136
probability of successful
        transmission, 33, 34
process noise, 46
producing process, 1
proximity method, 17
pseudo-Poisson path, 36
pulse-shaping filter, 105

quadratic error, 134
quantum efficiency, 9

radiation pattern, 6
radii, 41
random ray, 85
random walk, 135
range estimation, 20
ranging codes, 18
recursive solution, 45
reflective algorithm, 61, 62, 77, 93
reliability of the transmission, 33

resampling process, 50, 52
responsivity, 9, 14, 15
robotic applications, 39
robotic motion, 45
robots, 1
roof-antenna, 23
rotation bearings, 131
rotation sensors, 131

satellite infrastructures, 1
satellite signals, 3
scene analysis, 17, 21, 23, 27
semi-angle, 6, 13
semiconductor, 5, 6, 130
sensor fusion, xiii, 129, 130, 136–142
sensor techniques, 24
sensory data, 129
sensory feedback, 45
sequential estimate, 45
service discovery protocol, 23
shadowing, 121
shopping malls, 1
short transmission range, 5
shot noise, 14, 19, 91, 111, 116
signal processing, 45
signal propagation, 17
silicon micro-machining techniques,
    135
single carrier modulation, 105, 116
singularity in matrix, 98
site survey, 21, 26, 27
smart computational algorithms, 129
smart phones, xiii, 21, 129
social networks, 1
soft sensors, 129
solid angle, 6, 8
spacecraft, 26, 45
spectral efficiency, 105
square layout, 31
stable platform systems, 131, 133
standard deviation, 135
state transition vector, 46
statistical noise, 45
statistics and control theory, 45
stereoscopic vision, 129

strap-down systems, 131, 133
subcarrier signals, 105
sum of the square error, 35
sustainability, 130
synchronization, 14, 18, 20, 27, 31,
    42, 105, 114
system model, 5, 31, 32, 84, 111

target coordinates, 51, 54
temporally stable, 24
temporary service outage, 130
tension changes, 134
thermal noise, 14, 15, 91, 111, 116
thermo-mechanical white noise, 134,
    135
Thorlabs, Inc., 9
three-dimensional positioning, 61
time consumption, 1, 59
time delay, 19, 45
time domain, 79, 80, 110, 111
time series analysis, 45
time slots, 31, 33
time stamp, 18, 19
time synchronization, 105
time update phase, 47, 142
tolerance of deviation, 136
torque motors, 131
training sequence, 105, 114
training symbols, 114
trajectory optimization, 45
transimpedance receiver, 15
translational motion, 134
transmitted identification (ID) data,
    35
transmitted power, 13, 34, 81
transmitter, 5, 6, 13, 16, 19–23, 25,
    27, 31, 33–35, 42, 64, 65,
    67–72, 74–77, 81–84, 86,
    87, 89, 90, 98, 106, 111,
    112, 114, 115
transmitter's, 42
travelling time, 81
triangulation, 17
trilateration, 40
trilateration approach, 39, 68

trolley, 135
trust region algorithm, 61, 62, 98, 100
trusted region, 61
turntable, 135
two-phase positioning algorithm, 21

ubiquitous communications, 131
ubiquitous feature, 130
ultra-white, 6
uncertainty reduction, 129
unconstrained problem, 61
updated commands, 45
urban populations, 1

velocity, xiii, 25, 26, 48, 129, 131, 135

vertical RMS error, 64
vertical variations, 39
vertical view, 41, 42, 56
vibrating element, 134
viewing angle, 6
visible light, 5, 8, 16, 22, 31, 130

wavelength, 5, 9, 27
wearable devices, xiii, 129
weather conditions, 130
Weibull distribution, 23
wireless access power consumption, 130

yelp, 1

Zigbee, 3, 22, 26–28

Printed and bound by CPI Group (UK) Ltd, Croydon, CR0 4YY

17/10/2024

01775660-0020